MW01674369

Non-Conventional Hybrid Machining Processes

Manufacturing Design and Technology Series

Series Editor:
J. Paulo Davim
University of Aveiro, Portugal

This series will publish high-quality references and advanced textbooks in the broad area of manufacturing design and technology, with a special focus on sustainability in manufacturing. Books in this series should find a balance between academic research and industrial application. This series targets academics and practicing engineers working on topics in materials science, mechanical engineering, industrial engineering, systems engineering, and environmental engineering as related to manufacturing systems, as well as professions in manufacturing design.

Enabling Technologies for the Successful Deployment of Industry 4.0
Edited by Antonio Sartal, Diego Carou, and J. Paulo Davim

Industry 4.0: Challenges, Trends, and Solutions in Management and Engineering
Edited by Carolina Machado and J. Paulo Davim

Sustainable Manufacturing for Industry 4.0
An Augmented Approach
Edited by K. Jayakrishna, K.E.K. Vimal, S. Aravind Raj, Asela K. Kulatunga, M.T.H. Sultan, and J. Paulo Davim

Functional and Smart Materials
Edited by Chander Prakash, Sunpreet Singh, and J. Paulo Davim

Advanced Manufacturing and Processing Technology
Edited by Chander Prakash, Sunpreet Singh, and J. Paulo Davim

Characterization, Testing, Measurement, and Metrology
Edited by Chander Prakash, Sunpreet Singh, and J. Paulo Davim

Non-Conventional Hybrid Machining Processes
Theory and Practice
Edited by Rupinder Singh and J. Paulo Davim

For more information about this series, please visit: https://www.crcpress.com/Manufacturing-Design-and-Technology/book-series/CRCMANDESTEC

Non-Conventional Hybrid Machining Processes

Theory and Practice

Edited by

Rupinder Singh and J. Paulo Davim

CRC Press
Taylor & Francis Group
Boca Raton London New York

CRC Press is an imprint of the
Taylor & Francis Group, an informa business

First edition published 2021
by CRC Press
6000 Broken Sound Parkway NW, Suite 300, Boca Raton, FL 33487-2742

and by CRC Press
2 Park Square, Milton Park, Abingdon, Oxon, OX14 4RN

© 2021 Taylor & Francis Group, LLC

CRC Press is an imprint of Taylor & Francis Group, LLC

Reasonable efforts have been made to publish reliable data and information, but the author and publisher cannot assume responsibility for the validity of all materials or the consequences of their use. The authors and publishers have attempted to trace the copyright holders of all material reproduced in this publication and apologize to copyright holders if permission to publish in this form has not been obtained. If any copyright material has not been acknowledged please write and let us know so we may rectify in any future reprint.

Except as permitted under U.S. Copyright Law, no part of this book may be reprinted, reproduced, transmitted, or utilized in any form by any electronic, mechanical, or other means, now known or hereafter invented, including photocopying, microfilming, and recording, or in any information storage or retrieval system, without written permission from the publishers.

For permission to photocopy or use material electronically from this work, access www.copyright.com or contact the Copyright Clearance Center, Inc. (CCC), 222 Rosewood Drive, Danvers, MA 01923, 978-750-8400. For works that are not available on CCC please contact mpkbookspermissions@tandf.co.uk

Trademark notice: Product or corporate names may be trademarks or registered trademarks, and are used only for identification and explanation without intent to infringe.

Library of Congress Cataloging-in-Publication Data
Names: Singh, Rupinder, 1979- editor. | Davim, J. Paulo, editor.
Title: Non-conventional hybrid machining processes : theory and practice /
edited by Rupinder Singh and J. Paulo Davim.
Description: First edfition. | Boca Raton : CRC Press, 2020. |
Series: Manufacturing design & technology | Includes bibliographical references
and index.
Identifiers: LCCN 2020019571 (print) | LCCN 2020019572 (ebook) |
ISBN 9780367139131 (hbk) | ISBN 9780429029165 (ebk)
Subjects: LCSH: Machining.
Classification: LCC TJ1185. N576 2020 (print) | LCC TJ1185 (ebook) | DDC
671.3/5—dc23
LC record available at https://lccn.loc.gov/2020019571
LC ebook record available at https://lccn.loc.gov/2020019572

ISBN: 978-0-367-13913-1 (hbk)
ISBN: 978-0-429-02916-5 (ebk)

Typeset in Times
by codeMantra

Contents

Preface

Nowadays, the nonconventional machining processes (NCMPs) have become more or less conventional as these processes have virtually replaced the conventional material removal processes (CMRPs) at many places. But still one of the niche areas of CMRP lies in bulk material removal. Some developments in NCMP have helped to overcome the limitations (such as low material removal rate, high tool wear rate, and surface finish) in some of the defined NCMPs. In order to get more benefit from NCMP, the recent trend is on hybridization of different NCMPs and CMRPs. This book could act as a milestone for understanding the basic mechanism of NCMPs for their possible hybridization.

This book is focused on understanding the basic mechanism of some of the NCMPs for their possible hybridization (mainly for research and development). The master's/ PhD students can use this book for the development of basic framework on hybridization for the selected NCMP (for possible research thesis). The framework is further strengthened by case studies.

The concept of macro-modelling for hybrid NCMP and framework for the development of industrial standards has been outlined. Overall, this book covers the issues of process optimization and process capability for hybrid NCMP, and is probably one of the few books on combining NCMP and CMRP.

This book spreads in 9 chapters, and presents a comprehensive treatment of the process and equipment for hybrid NCMP. Each chapter is subdivided into subtopics and explains the subject matter in a systematic way.

SALIENT FEATURES

- Covers important NCMP from hybridization viewpoint.
- Provides case studies for process optimization of hybrid NCMP, especially for upcoming research areas.
- Maintains a balance of theoretical concept and mathematical analysis.

Rupinder Singh and J. Paulo Davim

Editors

Rupinder Singh received his PhD degree in Mechanical Engineering in 2006 from T.I.E.T, Patiala; Masters of Technology (Production Engineering) in 2001; and Bachelors of Technology (Production Engineering) in 1999 from Punjab Technical University. He is chartered engineer by Institution of Engineers (India) and UGC Research Awardee. Currently, he is professor at the Department of Mechanical Engineering, National Institute of Technical Teachers Training and Research, Chandigarh, India. He has more than 20 years of teaching and research experience in Production and Industrial Engineering, with special emphasis on additive manufacturing, nonconventional machining, and casting. He has guided large numbers of PhD and master's students, and coordinated 17 financed research projects. He has received several scientific awards. He has worked as evaluator of projects for international research agencies as well as examiner of PhD thesis for many Indian universities. He is the guest editor for several international journals, book editor, and advisor for many international conferences. In addition, he has also published as author (and co-author) for more than 20 monographs, 80 book chapters, and 500 articles in journals and conferences (more than 250 articles in journals indexed in SCI (SCOPUS/h-index:31+/3626+ citations)).

J. Paulo Davim received his PhD degree in Mechanical Engineering in 1997, MSc degree in Mechanical Engineering (Materials and Manufacturing Processes) in 1991, Mechanical Engineering degree (5 years) in 1986, from the University of Porto (FEUP), the Aggregate title (Full Habilitation) from the University of Coimbra in 2005, and the DSc (Higher Doctorate) from London Metropolitan University in 2013. He is senior chartered engineer by the Portuguese Institution of Engineers with an MBA and Specialist titles in Engineering and Industrial Management as well as in Metrology. He is also Eur Ing by FEANI-Brussels and Fellow (FIET) of IET London. Currently, he is professor at the Department of Mechanical Engineering of the University of Aveiro, Portugal. He is also distinguished as honorary professor in several universities/colleges. He has more than 30 years of teaching and research experience in Manufacturing, Materials, Mechanical, and Industrial Engineering, with special emphasis on machining and tribology. He has also interest in management, engineering education, and higher education for sustainability. He has guided large numbers of postdoc, PhD, and master's students, and has coordinated and participated in several financed research projects. He has received several scientific awards and honors. He has worked as evaluator of projects for ERC (European Research Council) and other international research agencies as well as examiner of PhD thesis for many universities in different countries. He is the editor in chief of several international journals, guest editor of journals, book editor, book series editor, and scientific advisory for many international journals and conferences. Presently, he is an editorial board member of 30 international journals and acts as a reviewer for more than 100 prestigious Web of Science journals. In addition, he has also published as editor (and co-editor) for more than 150 books and as author (and co-author) for more than 15 books, 100 book chapters, and 500 articles in journals and conferences (more than 250 articles in journals indexed in Web of Science core collection/h-index 55+/9500+ citations, SCOPUS/h-index 60+/12000+ citations, Google Scholar/h-index 77+/19500+ citations).

Contributors

Atul Babbar
Mechanical Engineering Department
Thapar Institute of Engineering and Technology, Thapar University
Patiala, India

Kamaljit Singh Boparai
Department of Mechanical Engineering
GZSCCET, MRS Punjab Technical University
Bathinda, Punjab, India

Gurinder Singh Brar
Department of Mechanical Engineering
National Institute of Technology,
Uttarakhand, India

Jasgurpreet Singh Chohan
Department of Mechanical Engineering
Chandigarh University
Mohali, Punjab, India

Dharmpal Deepak
Department of Mechanical Engineering
Punjabi University
Patiala, India

S. Devgan
Department of Mechanical Engineering
Khalsa College of Engineering & Technology
Amritsar, India

Vishwas Grover
Mechanical Engineering Department
Ajay Kumar Garg Engineering College
Ghaziabad, India

Dheeraj Gupta
Mechanical Engineering Department
Thapar Institute of Engineering and Technology
Patiala, India

Vivek Jain
Mechanical Engineering Department
Thapar Institute of Engineering and Technology
Patiala, India

Nimo Singh Khundrakpam
Department of Mechanical Engineering
National Institute of Technology
Manipur, India

Sudhir Kumar
Dept of Mechanical Engineering
Thapar Institute of Engineering and Technology
Patiala, India

A. Mahajan
Department of Mechanical Engineering
Khalsa College of Engineering & Technology
Amritsar, India

Pulak M. Pandey
Mechanical Engineering Department
Indian Institute of Technology Delhi
New Delhi, India

Richa Rani
Department of Chemistry
Panjab University
Chandigarh, India

Ankit Sharma
Chitkara College of Applied Engineering
Chitkara University
Patiala, India

Sarabjeet Singh Sidhu
Mechanical Engineering Department
Beant College of Engineering and Technology
Gurdaspur, India

Gurminder Singh
Mechanical Engineering Department
Indian Institute of Technology Delhi
New Delhi, India

Gurpreet Singh
Mechanical Engineering Department
Beant College of Engineering and Technology
Gurdaspur, India

Rupinder Singh
Department of Mechanical Engineering
NITTTR
Chandigarh, India

Tarlochan Singh
Production and Industrial Engineering Department
Dr. B.R. Ambedkar National Institute of Technology
Jalandhar, India

1 Ultrasonic Vibration-Assisted Sintering

Gurminder Singh and Pulak M. Pandey
Indian Institute of Technology Delhi

CONTENTS

1.1 INTRODUCTION

1.1.1 ULTRASONIC

From the early decades of the 20th century, ultrasonic vibrations have been used to improve different types of manufacturing processes. Ultrasonic vibrations are the physical vibrations at the molecular level that are transferred through any medium (solid, liquid, or gas) [1]. Twenty kilohertz to several gigahertz is generally considered as the ultrasonic vibration range. The assistance of ultrasonic vibrations in the manufacturing processes not only improves the effectiveness of the process but also reduces the cost of the process in terms of energy, consumption, and maintenance [2,3]. Several works have been carried out in different manufacturing fields and revealed that the ultrasonic effect is similar to that of thermal softening. A similar kind of thermal softening requires 10^7 times of thermal energy compared to the ultrasonic vibrations [4,5]. Ultrasonic waves have been widely used in industries and research fields for different purposes such as ultrasonic welding, milling, drilling, and casting to minimize cutting forces, reduce residual stresses, and accelerate the rate of reactions [6–8].

1.1.2 Sintering

Sintering involves the combination of metal, ceramic, or other powders into their solid parts. From thousands of years, sintering techniques have been used to fabricate pottery, bricks, jewelry, and other materials [9]. It is defined as the diffusion of particle atoms of metal powder below the melting temperature (60%–80%) to reduce surface area and porosity, and to enhance mechanical, electrical, and thermal properties.

Several variants of sintering techniques have been developed in the last decade to improve the diffusion phenomena. One of the variants is pressure-assisted sintering, in which the mechanical pressure is used to increase the diffusion between the particles. Another variant is spark plasma sintering, in which the electrical assistance is employed to enhance the densification with external pressure and electrical spark [10]. This process has advantages such as lower sintering temperature, shorter holding time, and marked comparative improvements in properties of materials; its disadvantages are high cost, low production rate, etc., over the conventional sintering. In microwave sintering, the electromagnetic energy is used to increase the performance of the sintering process. The high-frequency alternating electromagnetic field changes the diffusion continuously and provides rapid heating for small loads [11]. Generally, it sinters one compact at a time and requires the particles size around the penetration depth of the microwave. Therefore, the acceleration in the sintering diffusion process by external energy is the active area of research.

In recent years, ultrasonic vibration has been incorporated in the sintering process to enhance the sintering diffusion process. The cyclic vibrations are used in different types of sintering process such as pressureless sintering, pressure-assisted sintering, and loose powder sintering. The vibrations accelerate the reactions and enhance the neck growth during the initial stage of sintering. In this chapter, the comprehensive review of the ultrasonic-assisted sintering (UAS) has been made. The horn design process of UAS has also been discussed. The contributing mechanisms in the UAS for the accelerated sintering rate have been thoroughly discussed. The significant processing parameters such as ultrasonic power, ultrasonic vibration time, sintering temperature, particles' size and shape, and pressure employed during sintering have been elaborated in detail. A case study is presented to study the ultrasonic vibration entry time by two different approaches. First, the ultrasonic vibrations were provided from the beginning of sintering time till 3, 6, 9, 10, 15, and 20 minutes. Later, the vibrations were provided after 9 and 18 minutes of soaking time. The entry time effect of ultrasonic vibrations on the relative density and sintered particles' morphology was studied. The challenges, advantages, and summary of this technique are presented in the last section of this chapter.

1.2 LITERATURE BASED ON ULTRASONIC-ASSISTED SINTERING

The comprehensive literature review related to UAS is presented in this section. Ultrasonic vibration has been used to sinter different types of materials with or without applying pressure. The improvement in properties of different types of materials by ultrasonic vibration with or without pressure sintering is given in Table 1.1. The detailed literature review regarding UAS is presented below.

Chachin and Sedyako [12] reported the effect of ultrasonic vibrations on the sintering of copper-compacted powder. The authors revealed that the ultrasonic vibrations during sintering promote the contacts between the particles in the compacted powder. The vibrations disintegrated the oxide layer on the powder contacts, which decreased the friction forces and resulted in the better densifications. The ultrasonically treated samples were found to have 7% higher hardness as compared to the conventional process.

Lehfeldt [13] investigated the effect of ultrasonic vibrations on the hot pressing of nickel, aluminum, tin, and iron powder. It was reported that the ultrasonic vibrations reduced the ram pressure to attain the given compaction rate and to densify the material more homogenously in order to obtain better density irrespective of material types.

Abedini et al. [23] investigated the effects of the ultrasonic vibration on hot pressing of aluminum and found higher relative density as compared to the conventional hot pressing. The ultrasonic vibration provided a cyclic motion between the powder contact points, and improved the densification, resulting in higher hardness value.

Wei et al. [24] studied the effect of the ultrasonic vibration on the solid-state reaction between Fe_2O_3 and CaO for the formation of $CaFe_2O_4$ (CF) by stainless steel-based ultrasonic horn. The results indicated that the ultrasonic treatment accelerated the reactions and lowered the formation temperature of CF by 50°C as compared to the conventional process.

Li et al. [25] analyzed the effect of ultrasonic vibrations on the joining of silicon chips on the copper substrate by sintering of nano-silver paste. They found an improvement in microstructure densification of the sintered silver with decreased porosity (i.e., from 23.83% to 12.83%). Trung et al. [26] reported higher relative density and improved mechanical properties of the samples prepared by UAS of Cu-CNT (carbon nanotubes) nanocomposites by hot-isostatic pressing as compared to the conventional method. They concluded that the ultrasonic vibration mixed the composites homogenously and reduced the sintering temperature and time.

Similar research has been carried out by Abedini et al. [27] to study the microstructure and flexural strength in ultrasonic hot powder compaction of Ti-6Al-4V. They found that as the temperature was increased with ultrasonic assistance, the pores of the sintered samples were found to be reduced, and high flexural strength was measured as compared to the conventional methods.

Wang et al. [28] reported the effect of ultrasonic vibrations during sintering of silver nanoparticles and studied the combined effect of vibration and pressure. They revealed that the ultrasonic vibrations redistributed the particles in a homogenous

TABLE 1.1
UAS Literature Based on the Types of Sintering, Working Materials, and Improvements Due to Ultrasonic Assistance

Sr. No.	Type	Working Material	Property Improvement by Ultrasonic Vibration Assistance	Remarks	Ref.
1	Pressure assisted	Bismuth telluride alloy ($Bi_2Te_{2.7}Se_{0.3}$)	Vickers hardness increased by 82.9% Flexural strength increased by 50.2% Figure of merit (ZT) increased by 16.4%	The ultrasonic vibrations refined the grain size and improved the density of grain boundaries	[14]
2	Pressure assisted	Titanium	Density increased by 7.53% Friction stress reduced by 88.52% Spring-back effect reduced by 8.8%	The vibration assistance showed better results for the fine powder compared to the coarse powder	[15]
3	Pressure assisted	Piezoceramic and copper	Hardness increased by 2 times Density increased by 7.7%	More significant effect of ultrasonic vibrations resulted in vacuum	[16]
4	Loose pressureless	Copper	Relative density increased by 8%–14% Hardness increased by 10%–28% Electrical conductivity increased by 9.8%–12%	The ultrasonic vibrations removed the oxide layers from powder and exhibited pure copper surface for necking	[17–19]
5	Pressure assisted	Copper + tin	The phase is transferred from Cu_6Sn_5 to Cu_3Sn	The Cu_3Sn joints consist of equiaxed grains in each direction	[20]
6	Pressure assisted	Copper + tin	Fast reduction of Cu_3Sn phase Better microstructure densification	The mixed large and small particles revealed low priority due to ultrasonic vibration	[21]
7	Pressureless	Bronze	Density increased by 7.8%–9.8% Porosity reduced by 8%–10.2%	The cyclic stress increased the sintering driving force and densified the material with small pores	[22]

manner and improved the effective contact area as compared to the conventional sintering process. The ultrasonic vibrations only affected the initial stage of sintering and showed no significant effect after 3 minutes on the resistivity of the sintered silver particles. The molecular simulation revealed that the combination of pressure and ultrasonic vibration produced the cyclic stress on the particles during sintering, which introduced more dislocation lines and densified the particles' microstructure.

1.3 HORN DESIGN FOR UAS

The basic elements of the UAS process are an ultrasonic generator, transducer, horn, and crucible or sample holder, as shown in Figure 1.1. The high-frequency current generator is used for the oscillatory energy production. The generator gives control over the power and the frequency of the oscillating waves. The main function of the generator is to amplify the low-frequency (50 Hz) electrical energy to ultrasonic frequency (≥20 kHz). Many machines have been produced with a power range of 5–15 kW or more at frequencies from 19 to 22 kHz or more with an error of 0.5% [29]. These generators are tunable, which can be used to tune the attached components such as transducers and horn. Some are auto-tunable, which match the output frequency to the exact resonant frequency of the attached transducer/horn assembly [1]. Some generators are having manual adjustments to match the frequency with horns [1].

Further, the amplified oscillating electrical energy is converted into mechanical energy using a transducer. As per their working principles, these transducers are generally of two types: piezoelectric [30–33] and magnetostrictive [1,34,35]. High electromechanical conversion competence is available in the piezoelectric transducers, which do not require a cooling system. These transducers are not damaged by heat and are mostly used in the modern manufacturing industry. The magnetostrictive transducers have low energy conversion efficiency and require a cooling system to eliminate waste heat. Although this type of transducer has low efficiency, it can work in high-frequency ranges (17–23 kHz). These transducers also give better flexibility to design different types of horns compared to piezoelectric transducers. Both types of transducers have been used in UAS process [12,17,22,23,27]. The vibrational motion generated by the transducer is low in amplitude and difficult for practical use.

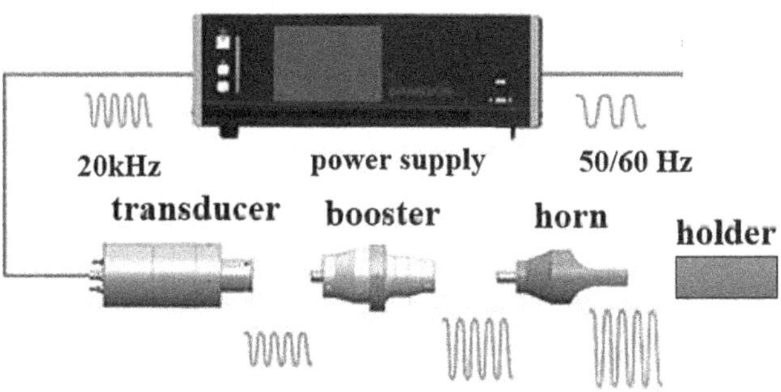

FIGURE 1.1 Ultrasonic system with different attachments.

Thus, the magnification or amplification of the motion needs to be carried out as per the required applications.

The horn is used to amplify the amplitude of the ultrasonic waves. The most important part in developing an ultrasonic system is to design the ultrasonic horn. The resonance between the horn and the transducer transfers the waves from one end to another end. Generally, the horns can be designed in three shapes: step, exponential, and conical [36]. These shapes have their own merits and demerits. The design selection of horn shape depends upon the application requirements. The step horn shape has a high magnification factor [37]. On the other hand, the exponential shape is used where less stress concentration is required [38]. The conical shape is generally used to fabricate small horns [39]. The schematic and magnification factors of these horns are shown in Figure 1.2. Generally, three types of alloy materials are used to fabricate horn, namely, aluminum alloy, stainless steel alloy, and titanium alloy [36]. The selection of the materials depends upon the application requirements and is also concerned with the cost of the setup. Sometimes, cylindrical shape horn is used for the displacement transmission. In that case, a booster is used to amplify the amplitude, which is transferred by the horn. The design of the horn is done by two methods: analytical and computational. Several authors [36,40–44] have used these methods for the horn design, which are described in detail in the literature. First, the length of the horn is calculated by the analytical calculations. Further, the minor changes in length are done by using FEM (finite element method) to obtain the required natural frequency. The modal analysis is used to find out the mode shape and natural frequency [45]. Furthermore, at the obtained natural frequency, the harmonic analysis is carried out to check the amplitude of displacement at the other end of the designed horn [46].

Normally, the horns are designed as per half-wavelength length calculation with the positions of one node and two antinodes. For different applications, horns can be designed with multiple wavelengths to increase the length with the similar amplification factor [38]. In UAS, multinodes and antinodes are suitable for horn design with material that depends upon the temperature. The ultrasonic horns used in the literature are given in Table 1.2. It has been observed that most of the horns were made up of stainless steel and titanium. As the major concern of the horn design in the UAS is its working temperature; both the aforementioned materials can sustain

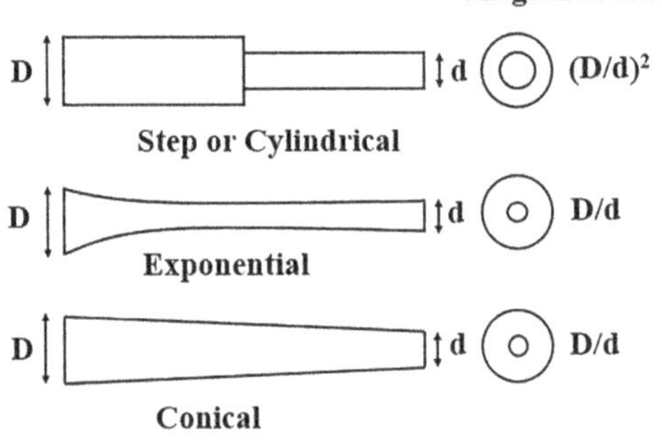

FIGURE 1.2 Different types of horn shapes with magnification factors.

TABLE 1.2
Horn Specifications Used in the Literature

Sr. No.	Horn Shape	Horn Material	Power, Frequency, and Amplitude	Working Temperature	Ref.
1	Step	Stainless steel	2kW, 20kHz	650°C -850°C	[24]
2	—	Steel	2kW, 20kHz, 5 μm	600°C–800°C	[12]
3	Conical	Titanium alloy	2kW,25kHz	400°C–600°C	[23,27]
5	Cylindrical	—	210 W, 40kHz, 12 μm	120°C	[28]
6	Cylindrical	Titanium alloy	210 W, 35kHz, 89%	350°C	[21]
7	Step	Stainless steel	2kW, 20kHz, 20 μm	500°C–900°C	[17,22]

their properties at high temperatures. On the other hand, small-sized horn designs are not preferred for UAS. Most pieces of the UAS research have been carried out using multiwavelength-designed horns with long length.

The crucible or sample holder design is also a precious task in UAS. As the maximum amplitude vibrations reach at the end of the tip of the horn, the sample is required to tightly attach to the horn end. For different types of sintering processes, different types of sample holders are designed. Abedini et al. [23,27] used a graphite-based sample holder for the hot pressing of the powders. The samples were placed on the tip of the horn to observe the maximum vibrations. The ultrasonic vibration also reduced the friction between graphite die and samples. A similar type of sample holders or crucibles were used for the UAS by different researchers [24,25,47]. For pressureless sintering, Singh and Pandey [18,19,22,48] have designed a steel crucible of the same length as the single-wavelength length, so that the maximum vibrations could be reached up to the loose powder particles.

1.4 WORKING PRINCIPLE

The ultrasonic energy provides cyclic motion to the particles in the UAS. This motion results in softening of the materials, which leads to high densifications. The working principle of the UAS depends upon the relation of heat energy with the ultrasonic energy. The reasoning of the UAS mechanism is concluded in the following points:

- Homogenous redistribution of the metal powder particles [17,25,28].
- Increase in local temperature between the powder particles [12,23,27].
- Decrease in material flow stress between the particles [23,47].
- Removal of moisture due to cyclic rubbing between particles [12,49].

The ultrasonic vibrations were transmitted to the particles via ultrasonic horn and solid particles by rarefaction and compression [24]. The schematic of the rarefaction and compression between the loose powder particles is shown in Figure 1.3. The given phenomena evenly redistribute the particles and increase the effective contact area for the necking as compared to the conventional sintering [28].

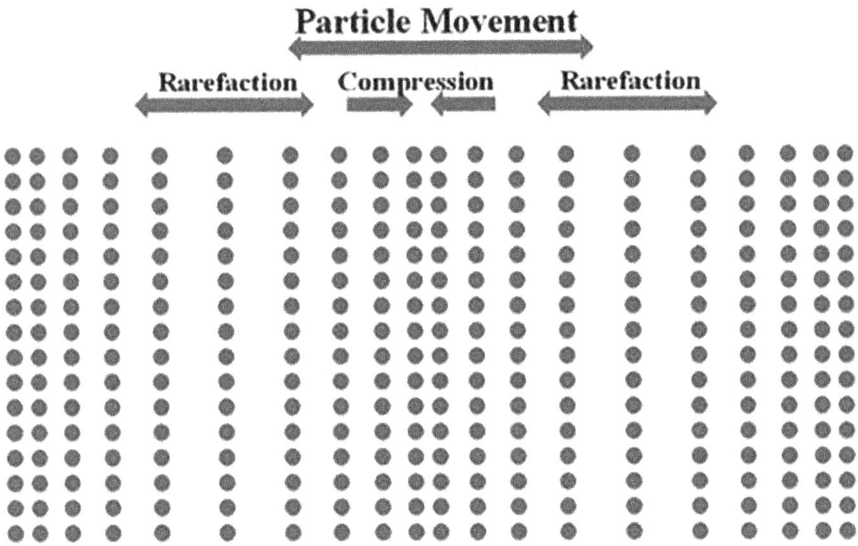

FIGURE 1.3 Schematic representation of rarefaction and compression between particles due to ultrasonic vibrations [24]

The schematic comparison between ultrasonic sintering and conventional sintering with sintering stages is shown in Figure 1.4. The particles that are sintered using UAS have uniform distribution and high diffusion as compared to the particles that are sintered using the conventional methods. The sufficient channels for the diffusion of the particles by losing surface energy will lead to the highly densified structure. The other reason for the increment in properties might be attributed to the decrement in material flow stress named as acoustic softening. The actual mechanism to decrease material flow stress involves the localized heating of the material around the dislocation using the scattering of ultrasonic waves [5]. It is believed that the ultrasonic vibration produces local heating between the particles' contact and the heat that is generated is absorbed by the particles to decrease the material

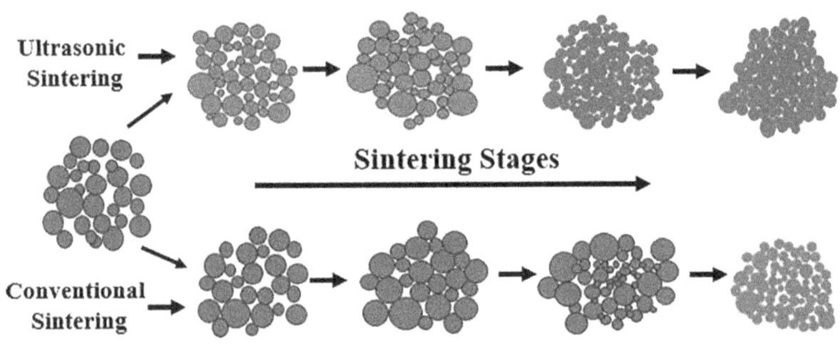

FIGURE 1.4 Schematic representation for the comparison of ultrasonic and conventional sintering.

flow stress [23]. The soft particles easily deform into the neighbor pores as compared to the hard particles and increase the rate of sintering to acquire highly densified structure. Moreover, for the proof of concept, the increase in local temperature was measured by different researchers at room temperature using thermocouple arrangements [22,23]. The researchers have found that the local temperature between the particles increased, i.e., double that of the room temperature. The higher ultrasonic power intensity resulted in a higher increase in local temperature. Hence, as the sintering temperature increased, the porosity of the sample was found to be reduced by losing the surface energy of the particles by the combined effect of sintering temperature and ultrasonic vibrations. The mechanical properties of the samples were found to be enhanced as the porosity decreased upon sintering with ultrasonic vibration. The ultrasonic waves remove the oxide layer by interparticle rubbing and expose the pure metal surface to contact [49]. The pure metal surfaces provided larger neck growth as compared to the oxidized metal surfaces.

1.5 IMPORTANT PROCESSING PARAMETERS

In UAS, there are different processing parameters that directly and indirectly influence the characteristics of the fabricated parts. The fish-bone diagram between processing parameters of UAS and performance characteristics is shown in Figure 1.5. The effect of different parameters is summarized in detail below.

1.5.1 Ultrasonic Power

Ultrasonic power is a crucial parameter in the UAS. The ultrasonic power percentage is directly proportional to the amplitude of the ultrasonic vibration. The increase in ultrasonic power percentage directly increases the transducer output vibration, which is further amplified using horn [37]. The ultrasonic vibrations reduce the material flow stress and increase the diffusion rate of sintering. The higher ultrasonic power

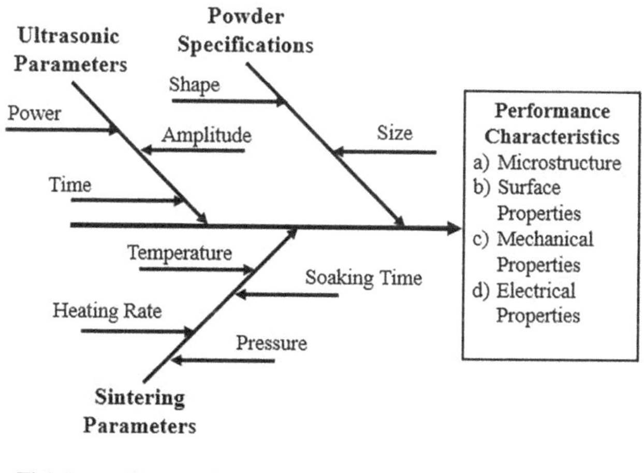

FIGURE 1.5 Fish-bone diagram between processing parameters and performance characteristics of UAS.

results in better diffusion as compared to lower ultrasonic power. Li et al. [25] studied the effect of ultrasonic power variations with respect to ultrasonic time on the shear strength of the nano-silver-sintered joint. They revealed that a slight increase in ultrasonic power gradually increased the shear strength (ref. Figure 1.6a).

A similar effect of ultrasonic vibrations on the shrinkage of compact copper sintering was noted by Pines et al. [47]. The shrinkage of the specimen increased with respect to ultrasonic power (vibration amplitude) due to the rapid sintering diffusion. The effect of ultrasonic power percentage on the chemical reaction between Fe_2O_3 and CaO was also studied [24]. The increase in ultrasonic power provided more energy, which accelerated the mass transfer and increased the formation of $CaFe_2O_4$ (ref. Figure 1.6b).

1.5.2 Ultrasonic Time

The ultrasonic vibrations enhanced the neck growth during the initial stage of sintering. Therefore, ultrasonic vibration time during sintering also plays a significant role in the properties of the sintered specimens. Li et al. [25] noticed that ultrasonic vibrations made a significant effect on the shear strength of sintered silver joint within 60 seconds (ref. Figure 1.7a). A similar trend of the ultrasonic time was noticed by Mei et al. [14] on the hardness of bismuth telluride alloy. No significant effect was observed after 10 minutes of the ultrasonic treating time. Chachin and Sedyako [12] discussed the effect of ultrasonic vibration time on the ultimate tensile strength (UTS) of the sintered specimen (at 600°C and 800°C). It was noticed that irrespective of sintering temperature, ultrasonic vibration showed no significant effect after 4 minutes of sintering time (ref. Figure 1.7b).

1.5.3 Sintering Parameters

There are several parameters that affect the properties of the specimen during the sintering process, such as heating rate, sintering time (soaking time), sintering temperature, cooling rate, sintering atmosphere, and compacting pressure. The effect

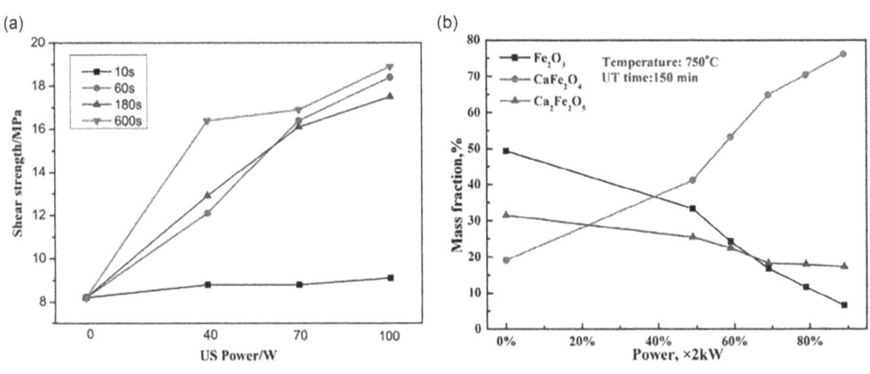

FIGURE 1.6 Effect of ultrasonic power (a) on the shear strength of sintered silver particles [25] and (b) chemical reaction between Fe_2O_3 and CaO [24].

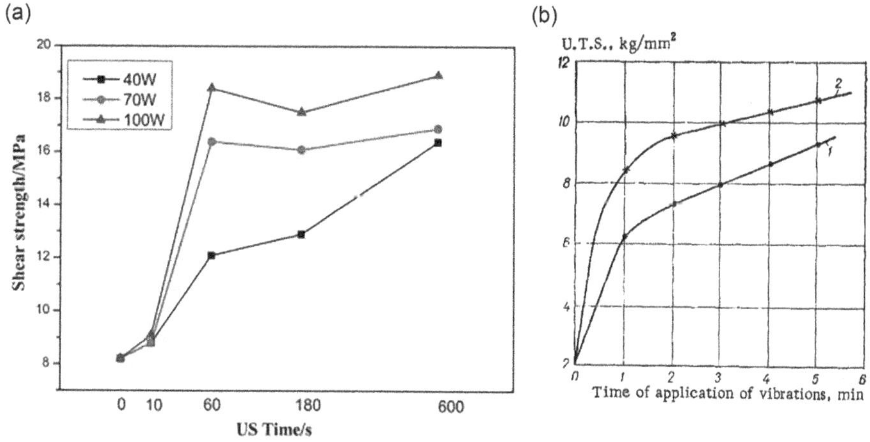

FIGURE 1.7 Effect of ultrasonic time on the (a) shear strength of sintered silver particle joint [25] and (b) UTS of copper specimens at 600°C and 800°C [12].

of these parameters on the sintering process has been reviewed in detail by several authors [50–55]. In UAS, sintering temperature plays a more significant role as compared to other sintering parameters. Singh and Pandey [17] compared the conventional pressureless sintering (CPS) and ultrasonic-assisted pressureless sintering (UAPS) with respect to the increase in sintering temperature. It was noticed that the relative density of the sintered part increased more rapidly with sintering temperature assisted by ultrasonic treatment (i.e., UAPS) as compared to CPS (ref. Figure 1.8a). The ultrasonic vibrations increased the local temperature between the particles due to the friction generated by vibrations. The combination of heat by sintering temperature and local temperature between the particles enhanced the

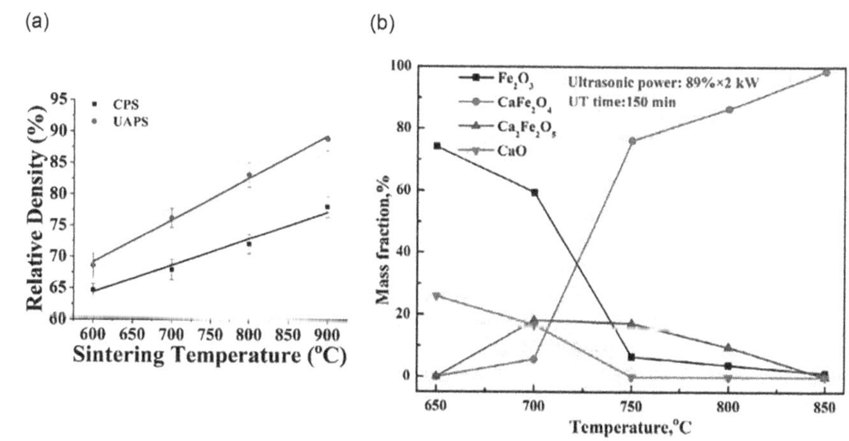

FIGURE 1.8 Effect of sintering temperature with vibration assistance on the (a) relative density of sintered copper [17] and (b) chemical reaction between Fe_2O_3 and CaO [24].

sintering process and resulted in better properties. Wei et al. [24] discussed the effect of sintering temperature with ultrasonic vibration assistance on the reaction between Fe_2O_3 and CaO. It was revealed that the combined effect of vibration and sintering temperature increased the mass fraction percentage of $CaFe_2O_4$ formation (ref. Figure 1.8b). The effect of sintering temperature in UAS is also discussed by other authors [12,14,23,26,27,47].

1.5.4 Pressure

The pressure during UAS provides a cyclic load on the specimen at the micron level. The cyclic load increases the connecting area between the particles and reduces the material flow stress. Few studies [13,26–28] have included pressure as the parameter while studying UAS process. Abedini et al. [27] studied the effect of pressure variations with a constant ultrasonic vibration on the relative density of Ti-8Al-4V. The relative density increased with an increase in cyclic pressure over the specimen (ref. Figure 1.9).

1.5.5 Powder Size and Shape

In sintering, powder particle size has a significant effect on the properties of the sintered specimen [50]. Abedini et al. [23] investigated the ultrasonic hot compaction effect at different temperatures with fine (40 μm) and coarse (250 μm) aluminum powder particle sizes. The ultrasonic vibration redistributed the fine particles more

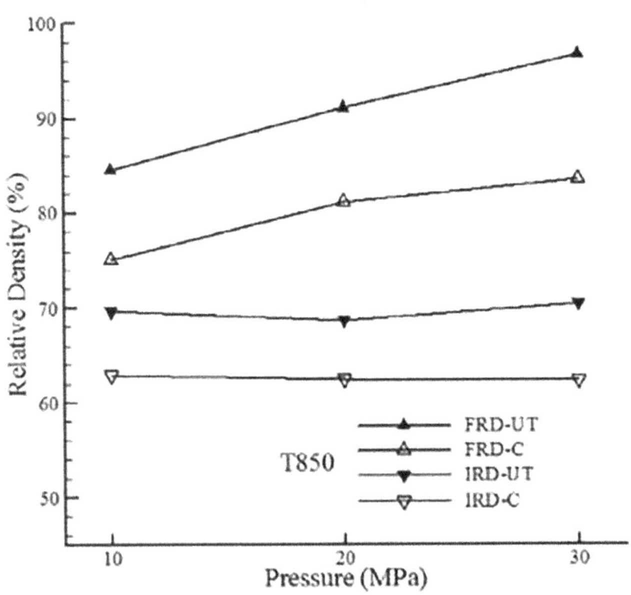

FIGURE 1.9 Effect of pressure with ultrasonic assistance on the relative density of Ti-6Al-4V [27].

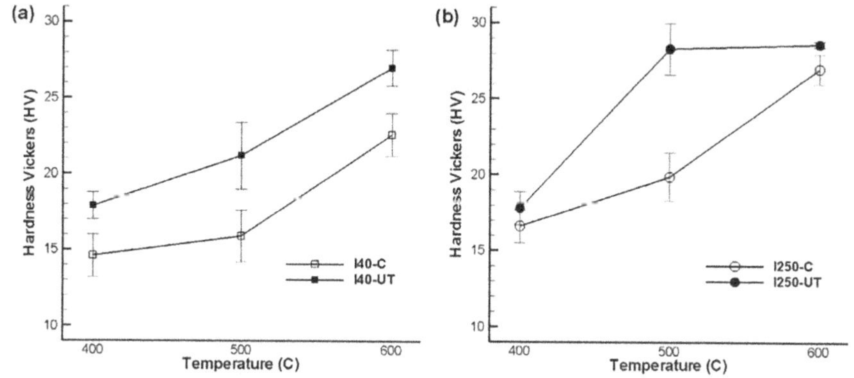

FIGURE 1.10 Ultrasonic treatment effect on the hot compaction of (a) 40 μm and (b) 250 μm irregular aluminum powder [23].

homogenously and significantly as compared to a coarse powder. The relative density was found to be 10% higher in ultrasonically treated fine particles as compared to coarse particles (ref. Figure 1.10). A similar redistribution of varying particle sizes was reported by Singh and Pandey [17].

1.6 EFFECT ON MICROSTRUCTURE

The ultrasonic vibrations softened the particles during UAS, leading to merging of small particles to the voids of large particles. The phenomena of cyclic stress between the particles lead to acoustic softening, thus showing a significant effect on the microstructure of the sintered specimens. Singh and Pandey [17] showed the effect of ultrasonic vibrations on the pressureless sintering of the loose copper powder. They revealed that the smaller particles evenly merged into the larger particles, which became spherical particles in ultrasonically treated samples as compared to the conventionally sintered samples (ref. Figure 1.11). The isolated spherical particles

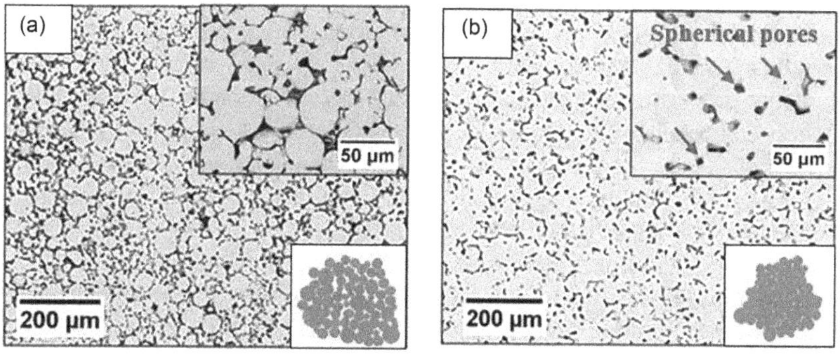

FIGURE 1.11 Microstructure of sintered copper loose powder using (a) conventional sintering process and (b) ultrasonic sintering process at 900°C sintering temperature [17].

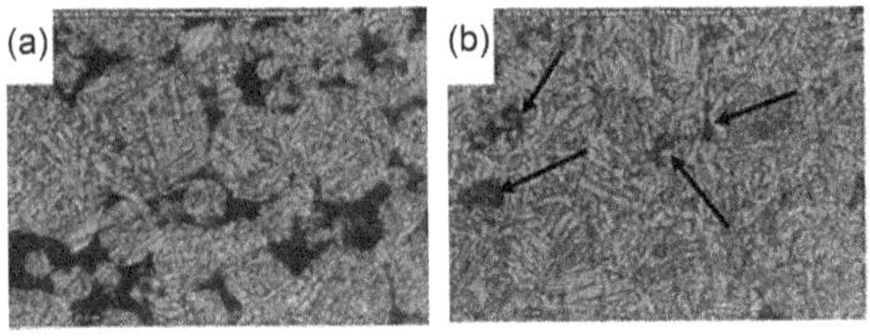

FIGURE 1.12 Microstructure comparison of pressure-assisted hot-compacted Ti-6Al-4V with (a) conventional process and (b) ultrasonic assistance at 850°C sintering temperature [27].

confirmed the final stage of sintering at 900°C sintering temperature with ultrasonic assistance [55]. On the other hand, high porosity and poor particle welding were observed in the conventional sintering process at 900°C sintering temperature. A similar ultrasonic vibration effect on the surface porosity and microstructure during the loose powder sintering was reported in the literature [24,25,56].

In pressure-assisted sintering, ultrasonic vibrations showed more significant effect on the microstructure of the sintered specimens as compared to UAPS. The ultrasonic vibrations with external ram pressure provided cyclic load stress on the particles in the pressure direction. The cyclic pressure forced the particles to diffuse into the neighboring particles in order to lower the surface porosity. Abedini et al. [27] reported the change in the microstructure of Ti-6Al-4V during ultrasonic-assisted hot compaction as compared to the conventional process. Low surface porosity and fine microstructure were observed in the ultrasonic vibration-treated samples as compared to conventionally fabricated samples (ref. Figure 1.12)

1.7 ADVANTAGES, CASE STUDY, CHALLENGES, AND FUTURE SCOPE

1.7.1 Advantages

The advantages of the UAS are summarized in the following points:

- Ultrasonic vibration accelerates the sintering diffusion process during the initial stage of sintering.
- Better properties such as relative density, electrical conductivity, and hardness can be achieved at low sintering temperature as compared to the conventional sintering process.
- Ultrasonic vibration enhanced the number of dislocations during sintering and resulted in the fine microstructure.
- Ultrasonic vibration enhanced the chemical reactions during sintering at a faster rate.
- Ultrasonic vibration homogenously mixed the metal powder particles to lower the porosity in the sintered specimen.
- Ultrasonic vibrations removed the oxide layer on the powder particles by rubbing at room temperature and exposed pure metal surfaces for sintering.

1.7.2 Case Study

To check the efficacy of the ultrasonic vibration effect on the sintering, few experiments have been carried out as a case study. The pressureless loose sintering of copper spherical powder is carried out with 20 μm ultrasonic vibrations. The metal powder particles were procured from MEPCO, India, with a particle size of 5–50 μm. The details of the experimental setup and measurement techniques used for the case study were provided in the previous studies [17,22]. The concern of the case study is to study the effect of ultrasonic times on the density and microstructure of the sintered particles. The comparison of UAPS with CPS was also studied. The sintering was carried out at 800°C with 4°C/min heating and cooling rate with 45 minutes of the soaking time. Two different ultrasonic vibration experiments were carried out with the same sintering cycle. First, to check the ultrasonic vibration time, ultrasonic vibrations were provided for heating rate from 600°C and for different times (3, 6, 9, 12, 15, 18, and 21 minutes) of soaking time. The vibrations were kept off for the CPS and compared with UAPS in terms of relative density and SEM morphology.

Figure 1.13 represents the effect of the ultrasonic time on the relative density of the sintered part and the relative density difference with CPS. It was observed that after 15 minutes of soaking time, there was no significant effect of ultrasonic vibrations on the sintering. The 10% increase in relative density as compared to CPS was observed. During the initial stage of sintering, surface mechanism plays a significant role [50]. The ultrasonic vibrations increased the local temperature between the particles and reduced the material flow stress for better densification. Due to this, small particle merged into the large particles faster as compared to the conventional method and provided higher neck growth. Using the SEM, morphological images of the sintered particles are also shown in Figure 13. It was observed that the

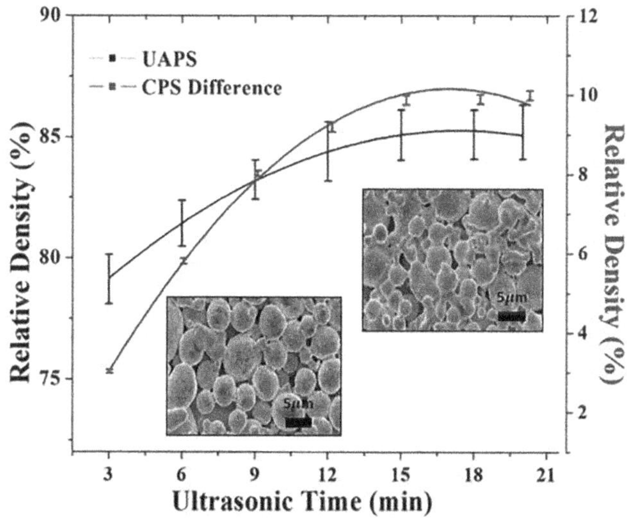

FIGURE 1.13 Effect of ultrasonic vibration time on the relative density of sintered copper with SEM images.

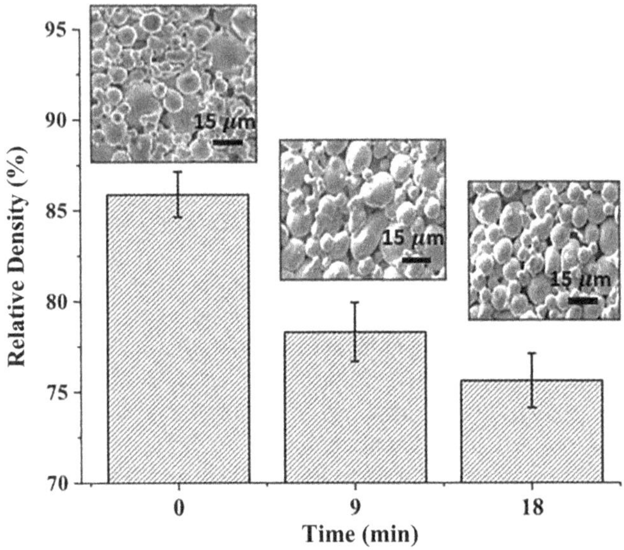

FIGURE 1.14 Effect of ultrasonic vibration entry time on the relative density of the sintered copper.

ultrasonically treated samples have larger necks between the particles as compared to the conventionally sintered particles. The grain-boundary diffusion mechanism was observed earlier in the case of UAPS as compared to CPS. Therefore, ultrasonic treatment during the initial stage of sintering accelerates the sintering mechanism and provides better densification at low temperature and less soaking time.

The next phase of experiments was conducted to check the efficacy of ultrasonic vibration's entry time. First, the vibrations were given from the starting of the soaking time (i.e., 0 minutes). Then, vibrations were given after 9 minutes of soaking time and later after 18 minutes of soaking time. These experiments were also compared based on the relative density and SEM morphological analysis. It has been observed that ultrasonic vibrations showed a significant effect from the starting of the soaking time. As the vibrations were given after 9 minutes of soaking time, 12% less relative density was observed as compared to the 0-minute entry results. Furthermore, after 18 minutes of soaking time, the ultrasonic vibrations showed an insignificant effect and the samples possessed 75% relative density similar to CPS samples. The SEM morphological results (ref. Figure 1.14) confirmed that the ultrasonic vibrations enhanced the neck growth in the zero-entry time-treated samples as compared to other samples.

1.7.3 Challenges and Future Scope

In the manufacturing processes, there are several types of challenges regarding design, parametric variations, characterizations, and repeatability. In UAS, the major challenge is to provide directly ultrasonic vibrations to the specimen due to the high-temperature treatment. Different means of sources have been discussed in the literature. For nanoparticles' sintering, direct transducer vibrations were used by the help of fixtures [25]. However, special attention was paid in the literature to the ultrasonic horn design and sintering accessories for the high-amplitude vibrations. The long-length horn design is still a challenge in UAS to prevent the damage of transducer and ultrasonic generator accessories from the furnace heat. The ultrasonic vibrations enhanced the initial stage of sintering in terms of large neck growth. The sintering conditions vary depending upon the type of materials, particle size and shape, and nature of sintering (pressureless, pressure-assisted, loose powder, etc.). Therefore, for such conditions, the effect of ultrasonic vibration times and vibration entry times will be different.

1.8 SUMMARY

This chapter represents the fundamental mechanism and important parameters of the UAS. Ultrasonic vibrations reduced the material flow stress and increased the local temperature between the particles for homogenous diffusion. The horn design for the UAS process has been discussed in detail. Further, the working principle of the UAS was explained. The ultrasonic vibrations at room temperature removed the oxide layer by interparticle rubbing and exposed the pure metal surface to contact. The pure surfaces softened at high temperature using ultrasonic vibrations, thus leading to better surface diffusion. The important processing parameters such as ultrasonic power, ultrasonic vibration time, sintering temperature, particles' shape and size, and pressure during sintering have been discussed. The case study regarding the sintering of copper particle was also given. It was revealed that the ultrasonic vibration entry has a significant effect on the diffusion of the particles. The advantages, major challenges, and future scope have been provided in the last segment of this chapter.

REFERENCES

1. Kumar S, Wu CS, Padhy GK, Ding W. Application of ultrasonic vibrations in welding and metal processing : A status review. *Journal of Manufacturing Processes* 2017;26:295–322. doi:10.1016/j.jmapro.2017.02.027.
2. Kumar J. Ultrasonic machining—a comprehensive. *Machning Science and Technology* 2013;17:325–79. doi:10.1080/10910344.2013.806093.
3. Thoe TB, Aspinwall DK, Wise MLH. Review on ultrasonic machining. *International Journal of Machine Tools and Manufacture* 1998;38:239–55.
4. Yadav S, Doumanidis C. Thermomechanical analysis of an ultrasonic rapid manufacturing (URM) system. *Journal of Manufacturing Processes* 2005;7:153–61.
5. Langenecker B. Effects of ultrasound on deformation characteristics of metals. *IEEE Transactions on Sonics and Utrasonic* 1966;13:1–8.
6. Singh R, Khamba JS. Comparison of slurry effect on machining characteristics of titanium in ultrasonic drilling. *Journal of Materials Processing Technology* 2007;7:200–5. doi:10.1016/j.jmatprotec.2007.06.026.
7. Shen N, Samanta A, Ding H, Cai WW. Simulating microstructure evolution of ultrasonic welding of battery tabs. *Procedia Manufacturing* 2016;5:399–416. doi:10.1016/j.promfg.2016.08.034.

8. Liang G, Shi C, Zhou Y-jun, Mao Da-heng. Numerical simulation and experimental study of an ultrasonic waveguide for ultrasonic casting of 35CrMo steel. *Journal of Iron and Steel Research International* 2016;23:772–7. doi:10.1016/S1006–706X(16)30119-4.
9. Fedorchenko IM, Skorokhod VV. Theory and practice of sintering. *Soviet Powder Metallurgy and Metal Ceramics* 1967;6:790–805. doi:10.1007/BF00773720.
10. Wang X, Casolco SR, Xu G, Garay JE. Finite element modeling of electric current-activated sintering: The effect of coupled electrical potential, temperature and stress. *Acta Materialia* 2007;55:3611–22. doi:10.1016/j.actamat.2007.02.022.
11. Demirskyi D, Agrawal D, Ragulya A. Neck growth kinetics during microwave sintering of copper. *Scripta Materialia* 2010;62:552–5. doi:10.1016/j.scriptamat.2009.12.036.
12. Chachin VN, Sedyako GK. Effects of ultrasonic vibrations on the sintering of metal-powder materials. *Soviet Powder Metallurgy and Metal Ceramics* 1968;7:693–4.
13. Lehfeldt E. The effect of ultrasonic vibrations on the compacting of metal powders. *Ultrasonics* 1967;5:219–23. doi:10.1016/0041–624X(67)90065–0.
14. Mei D, Wang H, Yao Z, Li Y. Ultrasonic-assisted hot pressing of Bi_2Te_3-based thermoelectric materials. *Materials Science in Semiconductor Processing* 2018;87:126–33. doi:10.1016/j.mssp.2018.07.019.
15. Fartashvand V, Abdullah A, Ali Sadough Vanini S. Effects of high power ultrasonic vibration on the cold compaction of titanium. *Ultrasonics Sonochemistry* 2017;36:155–61. doi:10.1016/j.ultsonch.2016.11.017.
16. Tsujino J, Ueoka T, Suzuki H, Shinuchi S, Hashimoto K. Ultrasonic vibration press of metal and ceramics powder using complex vibration and vacuum condition. *IEEE Ultrasonics Symposium* 1991:973–8.
17. Singh G, Pandey PM. Ultrasonic assisted pressureless sintering for rapid manufacturing of complex copper components. *Materials Letters* 2019;236:276–80. doi:10.1016/j.matlet.2018.10.123.
18. Singh G, Pandey PM. Uniform and graded copper open cell ordered foams fabricated by rapid manufacturing: surface morphology, mechanical properties and energy absorption capacity. *Materials Science & Engineering A* 2019;761:138035. doi:10.1016/j.msea.2019.138035.
19. Singh G, Pandey PM. Rapid manufacturing of copper components using 3D printing and ultrasonic assisted pressureless sintering: experimental investigations and process optimization. *Journal of Manufacturing Processes* 2019;43:253–69. doi:10.1016/j.jmapro.2019.05.010.
20. Zhao HY, Liu JH, Li ZL, Zhao YX, Niu HW, Song XG, et al. Non-interfacial growth of Cu_3Sn in Cu/Sn/Cu joints during ultrasonic-assisted transient liquid phase soldering process. *Materials Letters* 2017;186:283–8. doi:10.1016/j.matlet.2016.10.017.
21. Pan H, Huang J, Ji H, Li M. Enhancing the solid/liquid interfacial metallurgical reaction of Sn + Cu composite solder by ultrasonic-assisted chip attachment. *Journal of Alloys and Compounds* 2019;784:603–10. doi:10.1016/j.jallcom.2019.01.090.
22. Singh G, Pandey PM. Design and analysis of long-stepped horn for ultrasonic assisted sintering. *21st International Conference on Advances in Materials and Processing Technology (AMPT)*, Dublin, Ireland: 2018.
23. Abedini R, Abdullah A, Alizadeh Y. Ultrasonic assisted hot metal powder compaction. *Ultrasonics Sonochemistry* 2017;38:704–10. doi:10.1016/j.ultsonch.2016.09.025.
24. Wei R, Lv X, Yang M, Xu J. Effect of ultrasonic vibration treatment on solid-state reactions between $Fe_2O_3$and CaO. *Ultrasonics Sonochemistry* 2017;38:281–8. doi:10.1016/j.ultsonch.2017.03.023.
25. Li Y, Jing H, Han Y, Xu L, Lu G. Microstructure and joint properties of nano-silver paste by ultrasonic-assisted pressureless sintering. *Journal of Electronic Materials* 2016;45:3003–12. doi:10.1007/s11664-016-4394-8.

26. Trung TB, Phuong DD, Van Luan N, Vasilievich RV, Dmitrievich SA. Effect of ultrasonic-assisted compaction on density and hardness of cu-cnt nanocomposites sintered by capsule-free hot isostatic pressing. *Acta Metallurgica Slovaca* 2017;23:30–6. doi:10.12776/ams.v23i1.861.
27. Abedini R, Abdullah A, Alizadeh Y. Ultrasonic hot powder compaction of Ti-6Al-4V. *Ultrasonics Sonochemistry* 2017;37:640–7. doi:10.1016/j.ultsonch.2017.02.012.
28. Wang F, Nie N, He H, Tang Z, Chen Z, Zhu W. Ultrasonic-assisted sintering of silver nanoparticles for flexible electronics. *Journal of Physical Chemistry C* 2017;121:28515–9. doi:10.1021/acs.jpcc.7b09581.
29. Generator HP, Brown JA, Lockwood GR. Low-cost, high-performance pulse generator for ultrasound imaging. *IEEE Transactions on Ultrasonics, Ferroelectrics, and Frequency Control* 2002;49:848–51. doi:10.1109/TUFFC.2002.1009345.
30. Kanda T, Makino A, Ono T, Suzumori K. A micro ultrasonic motor using a micro-machined cylindrical bulk PZT transducer. *Sensors and Actuators* 2006;127:131–8. doi:10.1016/j.sna.2005.10.056.
31. Kanda T, Makino A, Suzumori K, Morita T, Kurosawa MK. A cylindrical micro ultrasonic motor using a micro-machined bulk piezoelectric transducer. *IEEE Ultrasonics Symposium* 2004;2:1298–301. doi:10.1109/ULTSYM.2004.1418028.
32. Voronina SÃ, Babitsky V. Autoresonant control strategies of loaded ultrasonic transducer for machining applications. *Journal of Sound and Vibration* 2008;313:395–417. doi:10.1016/j.jsv.2007.12.014.
33. Beecham D. Sputter machining of piezoelectric transducers. *Journal of Applied Physics* 2003;4357:1–6. doi:10.1063/1.1657198.
34. Claeyssen F, Lhermet N, Le Letty R, Bouchilloux P. Actuators, transducers and motors based on giant magnetostrictive materials. *Journal of Alloys and Compounds* 1997;258:61–73.
35. Cai W, Feng P, Zhang J, Wu Z, Yu D. Effect of temperature on the performance of a giant magnetostrictive ultrasonic transducer. *International Journal of Vibroengineering* 2016;18:1307–18.
36. Nad M. Ultrasonic horn design for ultrasonic machining technologies. *Applied and Computational Mechanics* 2010;4:79–88.
37. Sharma V, Pandey PM. Optimization of machining and vibration parameters for residual stresses minimization in ultrasonic assisted turning of 4340 hardened steel. *Ultrasonics* 2016;70:172–82. doi:10.1016/j.ultras.2016.05.001.
38. Ensminger D. Ultrasonics: data, equations, and their practical uses. 2009;10. doi:10.1017/CBO9781107415324.004.
39. Gupta V, Pandey PM. An in-vitro study of cutting force and torque during rotary ultrasonic bone drilling. *Proceedings of the Institution of Mechanical Engineers, Part B: Journal of Engineering Manufacture* 2016;232:1549–60. doi:10.1177/0954405416673115.
40. Singh RP, Singhal S, Singh RP, Singhal S. Rotary ultrasonic machining: A review rotary ultrasonic machining : A review. *Materials and Manufacturing Processes* 2016;31:1795–824. doi:10.1080/10426914.2016.1140188.
41. Technologies N, Alexandru E, Nanu S, Niculae P, Marinescu I. Study on ultrasonic stepped horn geometry design and FEM simulation. *Nonconventional Technologies Review* 2011:25–30.
42. Seah KHW, Wong YS, Lee LC. Design of tool holders for ultrasonic machining using FEM. *Journal of Materials Processing Technology* 1993;37:801–16.
43. Yadava V, Deoghare A. Design of horn for rotary ultrasonic machining using the finite element method. *International Journal of Advanced Manufacturing Technology* 2008;39:9–20. doi:10.1007/s00170-007-1193-7.

44. Amin SG, Youssef HA. Computer-aided design of acoustic horns for ultrasonic using finite-element analysis machining. *Journal of Materials Processing Technology* 1995;55:254–60.
45. Naseri R, Koohkan K, Ebrahimi M, Djavanroodi F, Ahmadian H. Horn design for ultrasonic vibration-aided equal channel angular pressing. *International Journal of Advanced Manufacturing Technology* 2017;90:1727–34. doi:10.1007/s00170-016-9517-0.
46. Chhabra AD, Kumar RV, Vundavilli PR, Surekha B. Design and analysis of higher order exponential horn profiles for ultrasonic machining. *Journal of Manufacturing Science and Production* 2016;16:13–9. doi:10.1515/jmsp-2015-0012.
47. Pines BY, Omelyanenko IF, Sirenko AF. Change in kinematics of sintering under the action of ultrasonic vibration. *Powder Metallurgy and Metal Ceramics* 1967;8:680–3.
48. Singh G, Pandey PM. Experimental investigations into mechanical and thermal properties of rapid manufactured copper parts. *Proceedings of the Institution of Mechanical Engineers, Part C: Journal of Mechanical Engineering Science* 2019;234(1):82–95.
49. Doumanidis CC. Nanomanufacturing of random branching material architectures. *Microelectronic Engineering* 2009;86:467–78. doi:10.1016/j.mee.2009.02.024.
50. German R. *Sintering: From Empirical Observations to Scientific Principles.* London: Butterworth-Heinemann 2014. doi:10.1016/C2012-0-00717-X.
51. Rockland JGR. The determination of the mechanism of sintering. *Acta Metallurgica* 1967;15:277–86. doi:10.1016/0001–6160(67)90203-9.
52. Olevsky EA, Tikare V, Garino T. Multi-scale study of sintering: A review. *Journal of the American Ceramic Society* 2006;89:1914–22. doi:10.1111/j.1551-2916.2006.01054.x.
53. Kuang X, Carotenuto G, Nicolais L. Review of ceramic sintering and suggestions on reducing sintering temperatures. *Advanced Performance Materials* 1997;4:257–74. doi:10.1023/A:1008621020555.
54. Ashby MF. A first report on sintering diagrams. *Acta Metallurgica* 1974;22:275–89. doi:10.1016/0001-6160(74)90167-9.
55. Kang S. L. *Sintering Densification, Grain Growth, and Microstructure.* London: Elsevier Butterworth-Heinemann Linacre House 2005:9–18. doi:10.1016/B978-075066385-4/50009-1.
56. Ji H, Li M. Mechanism of ultrasonic-assisted sintering of Cu@Ag NPs paste in air for high-temperature power device packaging. *Proceedings—Electronic Components and Technology Conference 2018*; 2018-May: 1270–5. doi:10.1109/ECTC.2018.00195.

2 Hybrid EDM with Three-Phase Dielectric Medium

Nimo Singh Khundrakpam
National Institute of Technology Manipur

Gurinder Singh Brar
National Institute of Technology Uttarakhand

Dharmpal Deepak
Punjabi University

CONTENTS

2.1 INTRODUCTION

In today scenario, the properties (strength, hardness, toughness, etc.) of the materials are improved due to the development in material science. Thus, the hardness of the tool higher than that of the workpiece is essential for the conventional machining. It is a very tough work to machine modern hard materials by the conventional machining process (CMP). Moreover, the CMP has limitations: manufacturing complex molds and dies' cavities, and noneconomic. As a result, many manufacturing industries replaced the CMP by the nonconventional machining process (NCMP). In NCMP, the tool and the workpiece are not in direct contact. Thus, NCMP has capable to machine the complex and hard materials with negligible tool wear and better surface finishing, but it has a low material removal rate. As a result, many investigators have focused on the nonconventional hybrid machining process (NCHMP), which hybridizes an NCMP with another NCMP or CMP. The word 'hybrid' means to combine the processes for a positive influence in their characteristics (Zhu et al. 2013). The objective of the hybrid machining process (HMP) is to attain the '1 + 1 = 3 effect', which means obtaining higher advantages of HMP from the simple summation of the individual advantages (Lauwers et al. 2014) by minimizing the limitations of the machining process, such as mechanical, chemical and thermal damages.

The word 'hybrid' means a combination of the processes for a large influence on the process (Lauwers, Klocke, and Klink 2010) or a combination of two (or more) machining processes for the positive removal of materials (Rajurkar et al. 1999). Broadly, HMPs are mainly classified into two types: (a) mixed or combined type in which two/more energy forms/mechanisms are directly involved in the removal of material and (b) assisted type, in which only single energy form/mechanism is directly involved in the removal of material, while other mechanisms are used only for assisting (Gupta, Jain, and Laubscher 2016). The necessity for the hybrid processes depends on the energy and resources identified (Treppe et al. 2011), and the demand of industry.

2.1.1 Electrical Discharge Machine (EDM)

Electrical discharge machining is an advanced NCMP in various fields for machining finishing parts of the aerospace, surgical instruments, automobiles, dies, and molds. It is independent of the hardness of workpiece due to the noncontact cutting force between the workpiece and the tool. In spite of its advantages, EDM has many limitations: the high tool wear rate (TWR) and the low surface finish (Zhao, Meng, and Wang 2002). However, the dielectric medium plays as an insulator and a coolant between the polarized tool and the workpiece to induce electric discharge. It also plays an important role in the machining processes of EDM such as erosion of material during the discharge, influencing the expansion of plasma channel, and removal of debris, and restating the discharge gap to its original state after the discharge. Thus, many researchers have proposed the new machining processes to minimize the limitations of EDM by replacing the single-phase liquid dielectric medium by another single-phase (gas for dry EDM) and two-phase (liquid–gas for near-dry EDM or liquid–powder for powder-mixed EDM) dielectric medium. Replacing liquid dielectric medium by a gas dielectric medium reduced the TWR close to zero (Kunieda, Yoshida, and Taniguchi 1997). However, it exhibits unstability and more arcing during machining; as a result, lower quality of the surface finish occurs. These limitations can be improved by using the two-phase dielectric medium (liquid–gas). But the two-phase dielectric medium has a higher thermal load on electrodes, resulting in higher TWR for longer run (Bai, Yang, and Zhang 2018). EDM with two-phase dielectric medium can minimize pollution, fire hazards, and explosions as compared to the EDM with single-phase (liquid) dielectric medium (Yang, Zuo, and Yu 2010). The powder-mixed electrical discharge machine (PMEDM) used different fine conductive powders such as graphite (Gr), silicon (Si), chromium (Cr), aluminum (Al), boron carbide (B_4C), silicon carbide, titanium oxide (TiO_2), and copper (Cu) that are mixed with the dielectric medium to assist the EDM, which is known to develop an assisted-type HMP. Mixing of electrically conductive powder in the dielectric liquid (Kansal, Singh, and Kumar 2007a) and the presence of liquid droplets in the powder in gas dielectric reduce the dielectric strength, which results in the easy removal of eroded particles, and reduce the short circuit and arc discharge, thus improving the working stability. The high transfer of the carbon element occurred from the workpiece to the tool, and vice versa; as a result, more amounts of hard carbide on the surface are formed, which helps in improving the microhardness of the machined surface (Kumar et al. 2015, 2017) and also helps in decreasing the electrode wear ratio (Road and Park 2016).

It is found that mixing of the Cr, Al, and SiC with dielectric medium can improve the MRR (Tzeng and Lee 2001). However, Cu powder has a high density, which causes an imbalance in mixing and is also present at the bottom of the mixture. As a result, Cu powder contribution is not significant in the machining process. However, they also suggested Cr powder for better MRR as compared to powder containing Al, SiC, and Cu. They revealed that the size, density, concentration, thermal conductivity, and electric conductivity of the powder mainly affect the machining performance.

Gr powder is commonly used in PMEDM due to its higher lubricity properties and good electric conductivity, which help to increase the MRR (Jeswani 1981). Moreover, mixing of Gr powder improves the surface finish, nearly the mirror-finish surface (Wong et al. 1998). Moreover, it improves microhardness of the surface (Singh, Kumar, and Singh 2015) and reduces the surface tensile stress (Talla, Gangopadhyay, and Biswas 2017). Moreover, the use of powders such as Ti, Al, TiC, and W improves the wear resistance and surface hardness (Marashi et al. 2016). It is experimentally found that the higher concentration of Al_2O_3 powder reduces the powder deposition on the workpiece (Patel, Thesiya, and Rajurkar 2018). On the other hand, the higher concentration of powder increases more chances of the arcing (Batish, Bhattacharya, and Kumar 2014), which helps to reduce the MRR. So, it is essential to know the proper amount of powder to be mixed with the dielectric medium for better machining performance.

2.1.2 Hybrid EDM with Three-Phase Dielectric Medium

Hybrid EDM with the three-phase dielectric medium commonly known as the power-mixed near-dry EDM (PMND-EDM) (Gao, Zhang, and Zhang 2009) is a newly invented assisted-type NCHMP, which directly uses thermal energy for removing material, and the conductive powder helps in assisting the machining process. A minimum-quality lubrication (MQL) system is used to form the three-phase (liquid–gas–solid) dielectric medium. The working principle of PMND-EDM is shown in Figure 2.1. Liquid dielectric is premixed with the powder

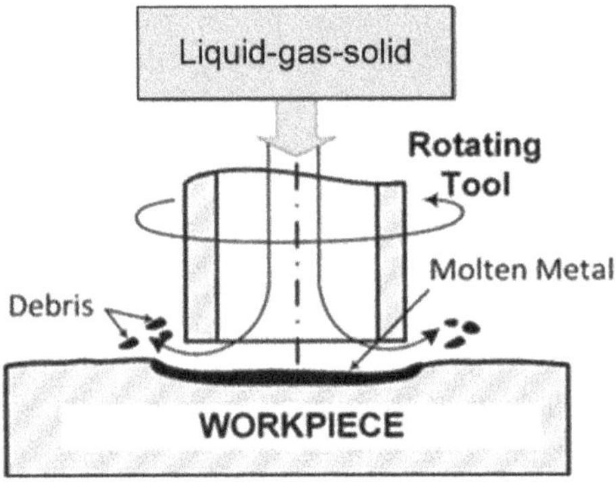

FIGURE 2.1 Principle of PMND-EDM.

additives and then mixed with the high-pressure gas to form the three-phase dielectric medium for PMND-EDM. Later, it is supplied at the interelectrode gap (IEG) between the tool and the workpiece. It acts as an insulator, a flushing agent, and a coolant to solidify the metal. Graphite, silicon, chromium, aluminum, boron carbide, and silicon carbide can be used as powder additives for the PMND-EDM process. The low machining efficiency and the high rate of short circuit are the common problems of dry EDM that can be effectively solved by PMND-EDM. The presence of powder material helps to enlarge the discharge gap, thus leading to easy removal of material from the discharge gap. It also helps in the removal of debris materials from the gap and prevents excessive heating of the tool and workpiece at the discharge spots.

In 2009, the feasibility of the PMND-EDM is suggested and revealed that mixing of conductive powder in the form of gas with liquid droplet dielectric medium reduced the dielectric strength of the EDM process (Gao, Zhang, and Zhang 2009). As a result, the larger discharging gap is formed at the IEG, in which eroded particles is sufficient to move (Gao, Zhang, and Zhang 2009). Thus, the arc discharging and the short circuit are reduced, which helps in improving the machining stability and better surface finish. They concluded that PMND-EDM has better surface finish due to swallowing of the electric erosion pit depth by the discharge dispersing the effect of powder. PMND-EDM has higher MRR as compared to dry EDM, which uses a single-phase gas dielectric medium (Bai et al. 2012a). Moreover, PMND-EDM has a higher discharge gap, thus improving the dispersed phase of discharge compared to dry EDM (Bai et al. 2012b). Hence, the limitations of dry EDM can be solved by minimizing the short circuit and improving the machining efficiency. The MRR of PMND-EDM increased with the increase in powder concentration, and also the TWR and SR also increased up to a certain value (Bai, Yang, and Zhang 2018). EDM with the three-phase dielectric medium (PMND-EDM) has better surface finish and negligible TWR as compared to the EDM. Thus, PMND-EDM is mainly preferable on the finishing process (Khundrakpam, Brar, and Deepak 2018).

In this study, hybrid EDM with three-phase (liquid–solid powder–gas) dielectric medium has been investigated on surface roughness and compared with the machined surfaces of EDM with the single-phase and two-phase dielectric medium. Later, optimization of the PMND-EDM process was carried out using the Taguchi method.

2.2 EXPERIMENTAL PROCEDURES

2.2.1 Experimental Setup

An experiment was conducted on EDM (make: Savita Machine Tools Pvt. Ltd, Pune, India). A MQL system (make: Dropsa) was used to produce and supply the two-phase (liquid–gas) and three-phase (liquid–gas–solid powder) dielectric mediums at the IEG, as shown Figure 2.2 (Khundrakpam, Brar, and Deepak 2018). The liquid dielectric medium was mixed at the MQL device with a gas dielectric medium to produce and supply the two-phase dielectric medium at IEG. Similarly, liquid dielectric medium and solid powder (<20 μm) were premixed at the MQL device with compressed gas to form and supply the three-phase dielectric medium at IEG. A solenoid valve is used to control the supply of compressed gas to the MQL device.

2.2.2 Materials

The experiment was conducted on EN-8 (0.40% C + 0.25% Si + 0.80% Mn + 0.015% S + 0.015% P) workpiece. The tool material was chosen as copper (99.9% purity) of diameter 10 mm with a hole of 2 mm diameter. In this experiment, three dielectric mediums of single-phase (deionized water), two-phase (deionized water–air), and three-phase (deionized water–air–graphite) were used to compare the finished surface of each EDM process. The deionized water and air are used as liquid and gas dielectric mediums, respectively, and the graphite powder (<20 μm) is used as a solid powder for this experiment.

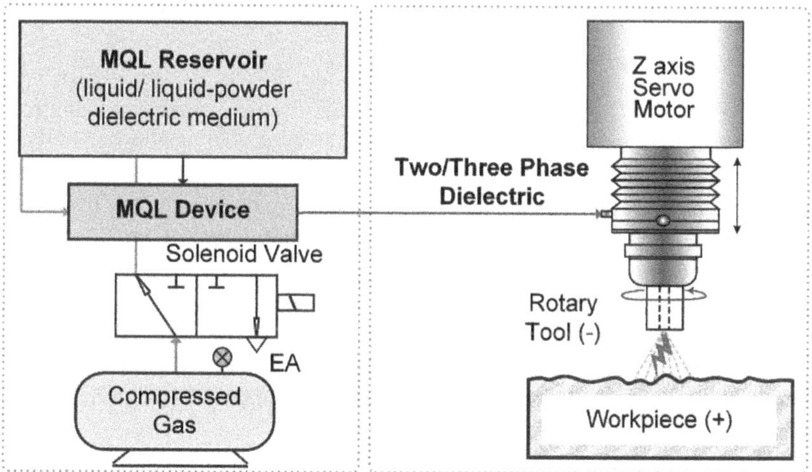

FIGURE 2.2 Working diagram of EDM with two-phase/three-phase dielectric medium.

2.2.3 Measuring Equipment

The Taylor Hobson CCI MP: Noncontact 3D optical profiler is used to measure the surface texture profile of machined surface. The 3D surface area roughness, S_a (μm), by using Eq. (2.1) and also the maximum height of surface, S_z (μm), are recorded. Later, the 2D surface profile is extracted diagonally and used to measure the 2D surface profile roughness; the R_a (μm) value of the machined surface is expressed in Eq. (2.2):

$$S_a = \frac{1}{A}\iint_A |H(x, y)| dxdy \tag{2.1}$$

where $H(x, y)$ is the height of the surface and A is the area measured.

$$R_a = \frac{1}{l}\int_0^l |H(x)|\, dx \tag{2.2}$$

where $H(x)$ is the height of the line profile within sampling length l; and S_z is the sum of the highest peak value and the deepest depth value within the area measured.

2.3 DESIGN OF EXPERIMENT (DOE)

2.3.1 DOE for Comparing the Surface Texture

In this experiment, the surface texture profiles of EDMed surfaces were first compared at different dielectric mediums, viz., single-phase (deionized water), two-phase (deionized water–air), and three-phase (deionized water–air–graphite) dielectric mediums. The machining processing parameters for comparing the surface texture of each EDM process were identified from the literature survey and machine capability (Table 2.1).

2.3.2 Taguchi Method

From the 1990s, the Taguchi method is a commonly used optimizing method for enhancing the productivity of R&D (Bendell, Disney, and Pridmore 1991). The Taguchi method is used for the single-objective analysis to determine the optimal processing parameters of each response. The Taguchi method is mainly used to optimize the single response. Initially, the processing parameters of MPND-EDM are determined, and appropriate Taguchi orthogonal array (OA) is selected for measuring the response. Then, the response is converted into the corresponding *S/N* ratio depending on the nature of problem in the form of 'larger-the-better' (LB), 'smaller-the-better' (SB), and 'nominal-the-better' (NB) characteristics as given in Eqs. (2.3), (2.4), and (2.5), respectively, and in the form of response decision matrix.

$$S/N_{larger\text{-}the\text{-}better} = -10\ \mathbf{log}_{10}\left[\frac{1}{n}\sum_{n=1}^{n}\frac{1}{y_i^2}\right] \tag{2.3}$$

TABLE 2.1
Machining Processing Parameters for the Experiment

Processing Parameter	Unit	Dielectric Phase of EDM: Single-Phase	Two-Phase	Three-Phase
Dielectric medium	—	Deionized water	Deionized water–air	Deionized water–air–graphite powder
Discharge current	A	5	5	5
Duty factor	%	0.8	0.8	0.8
Pulse on time	μs	100	100	100
Tool rotational speed	rpm	200	200	200
Powder concentration	g/L	—	—	8
Working pressure	bar	—	5	5
Tool polarity	—	Negative	Negative	Negative
Machining time	min	15	15	15

$$S/N_{smaller\text{-}the\text{-}better} = -10\ \mathbf{log}_{10}\left[\frac{1}{n}\sum_{n=1}^{n} y_i^{\,2}\right] \tag{2.4}$$

$$S/N_{nominal\text{-}the\text{-}better} = -10\ \mathbf{log}_{10}\left[\frac{\overline{y}_i^{\,2}}{\sigma^2}\right] \tag{2.5}$$

where y_i is the measured value of n number of repetitions; $\overline{y}_i$ and σ are the mean and standard deviation of the measured values.

In this experiment, Taguchi L_{18} OA is chosen for optimizing the processing parameters of PMND-EDM on the 3D surface area roughness, S_a (μm). The machining parameters for the experiment are given in Table 2.2.

TABLE 2.2
Processing Parameters of PMND-EDM

Parameter	Symbol	Units	Level
Tool polarity	—	—	Negative
Machining time	—	min	20
Working pressure	—	bar	5
A: Duty factor	DF	%	70, 80
B: Pulse on time	Ton	μs	50, 180, 320
C: Discharge current	DC	A	5, 10, 15
D: Tool rotational speed	N	rpm	50, 200, 350
E: Powder concentration	PC	g/L	0, 5, 10
F: Gap voltage	V	min	45, 50, 55

2.4 RESULTS AND DISCUSSION

2.4.1 Comparison of the Surface Texture of EDM Processes

The machined surfaces are cleaned with acetone, and the surface texture profile of the machined surface is measured using a noncontact 3D optical profiler. The ISO 25178 system is used to measure the S_a value and S_z value. Later, the measured surface texture profile is extracted diagonally, and the R_a value is measured with ISO 4287 system for a cutoff length of 0.08 mm. The 3D surface texture profile of EDMed surface with the single-phase (deionized water) dielectric medium is measured, as shown in Figure 2.3a. It is observed that 4.886 μm with a tolerance of ±0.489 *μm* is considered as S_a value and 89.705 μm as S_z value. Later, 2D surface profile is extracted from the measured surface diagonally (the upper-right to the lower-left corner), as shown in Figure 2.4a. Corresponding R_a value of the extracted surface is measured as 3.058 μm.

Figures 2.3b and 2.4b show the 3D surface texture profile and the extracted 2D surface profile of the EDMed surface with two-phase (deionized water–air) dielectric medium. The measured surface has the S_a value of 3.847 μm with a tolerance of ±0.385 μm, the S_z value of 56.621 μm, and the R_a value of 2.780 μm. The two-phase dielectric medium reduced the surface area roughness: the S_a by 1.165 μm, S_z by 33.084 μm, and R_a value by 0.278 μm. Simply, the use of the two-phase

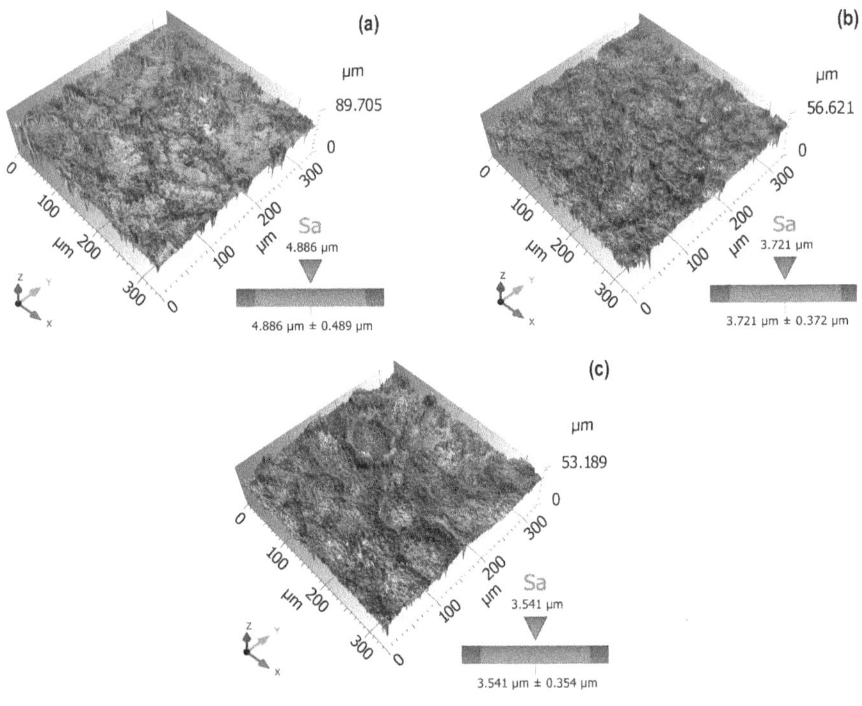

FIGURE 2.3 3D surface texture of EDM with (a) single-phase, (b) two-phase, and (c) three-phase dielectric mediums.

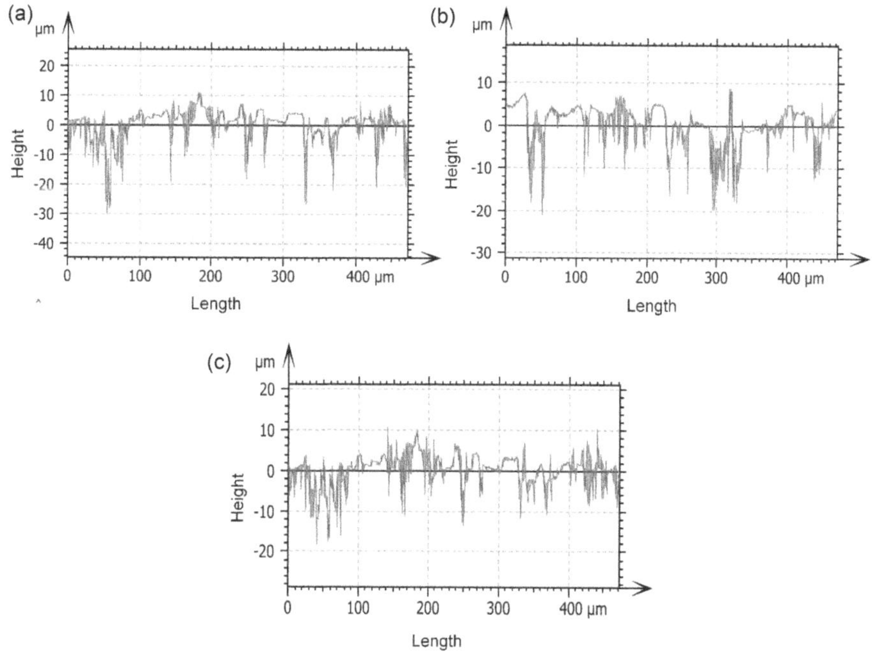

FIGURE 2.4 The extracted 2D surface profile of EDM with (a) single-phase, (b) two-phase, and (c) three-phase dielectric mediums.

dielectric medium in EDM process can reduce the surface roughness parameters of the machined surface by 31.3% on S_a, 58.4% on S_z, and 10% on R_a from the EDM with the single-phase dielectric medium. This is attributable to EDM with two-phase dielectric medium that has more cooling capacity, which can improve the deionization of plasma channel and thus results in stability in machining.

Figures 2.3c and 2.4c show the 3D surface profile and diagonally extracted 2D surface profile of the EDMed surface by using the three-phase (deionized water–air–graphite powder) dielectric medium. The measured surface has the S_a value of 3.627 µm with a tolerance of ±0.847 µm, the S_z value of 53.189 µm, and the R_a value of 2.647 µm. The three-phase dielectric medium in EDM process can reduce the surface roughness parameters: the S_a by 1.345 and 0.180 µm, the S_z by 36.516 and 3.432 µm, and the R_a by 0.560 and 0.220 µm from the single-phase and two-phase dielectric mediums. Simply, the three-phase dielectric medium in EDM process reduces the surface roughness parameters of the machined surface by 38.0% and 5.1% on S_a, 36.5% and 3.4% on S_z, and 15.5% and 5% on R_a from the EDM with single-phase and two-phase dielectric mediums, respectively. This is attributable to the effect of conductive powder mixing in the two-phase (liquid–gas) dielectric medium that reduces the dielectric strength and maintains larger discharging gap, resulting in improving machining stability, removing the eroded particles, and also reducing the short circuit. Moreover, dispersion of discharge due to the presence of powder reduces the erosion pit depth and widens the plasma channel, thus improving the surface quality. Moreover, the three-phase dielectric medium is flushing within the IEG that enhances the cooling capacity of both the electrodes, thus also improving the surface finish.

2.4.2 Optimization of Processing Parameters

The Taguchi L_{18} OA design matrix with the values of 3D surface area roughness, S_a (μm), is given in Table 2.3. In this study, minimizing the 3D surface area roughness, S_a, is primarily concerned. The raw data of 3D surface area roughness, S_a, has 'SB' characteristics, whereas the *S/N* ratio has 'LB' characteristics. Therefore, the variation trend of the means and the corresponding *S/N* ratios are opposite among them. The average value of S_a is depicted in Table 2.4. Then, the influence of processing parameters on the mean and the *S/N* ratios of S_a is shown in Figures 2.5 and 2.6, respectively.

From Figures 2.5 to 2.6, it is observed that the mean of S_a increases and the corresponding *S/N* ratio of S_a decreases as the value of discharge current, duty factor, and pulse on time increases. Thus, the rough surface finish is observed due to an increase in larger size and the deeper depth of crater on the machined surface with the higher value of pulse on time and discharge current. Moreover, the increase in discharge current is accompanied by the larger localized discharge energy with higher melting

TABLE 2.3
Taguchi L_{18} OA Design Matrix with Response Table

	Processing Parameters						3D Surface Area Roughness (S_a)	
Run	**A**	**B**	**C**	**D**	**E**	**F**	**Raw Data (μm)**	***S/N* Ratios (dB)**
1	1	1	1	1	1	1	2.598	−8.2938
2	1	1	2	2	2	2	2.779	−8.878
3	1	1	3	3	3	3	3.421	−10.683
4	1	2	1	1	2	2	2.496	−7.945
5	1	2	2	2	3	3	2.676	−8.549
6	1	2	3	3	1	1	3.874	−11.763
7	1	3	1	2	1	3	2.846	−9.085
8	1	3	2	3	2	1	3.234	−10.195
9	1	3	3	1	3	2	3.676	−11.308
10	2	1	1	3	3	2	2.489	−7.921
11	2	1	2	1	1	3	3.742	−11.462
12	2	1	3	2	2	1	4.130	−12.319
13	2	2	1	2	3	1	2.651	−8.468
14	2	2	2	3	1	2	3.720	−11.411
15	2	2	3	1	2	3	4.315	−12.699
16	2	3	1	3	2	3	3.163	−10.002
17	2	3	2	1	3	1	3.428	−10.701
18	2	3	3	2	1	2	4.496	−13.057

TABLE 2.4
Average Value of 3D Surface Area Roughness, S_a

	3D Surface Area Roughness (S_a)					
	Raw Data			S/N Ratio (dB)		
Processing Parameters	**Level 1**	**Level 2**	**Level 3**	**Level 1**	**Level 2**	**Level 3**
A: Duty factor	3.07	3.57	—	−9.633	−10.893	—
B: Pulse on time	3.19	3.29	3.47	−9.926	−10.139	−10.724
C: Discharge current	2.71	3.26	3.99	−8.619	−10.199	−11.971
D: Tool rotational speed	3.38	3.26	3.32	−10.401	−10.059	−10.329
E: Powder concentration	3.55	3.35	3.06	−10.845	−10.340	−9.605
F: Gap voltage	3.32	3.28	3.36	−10.290	−10.086	−10.414

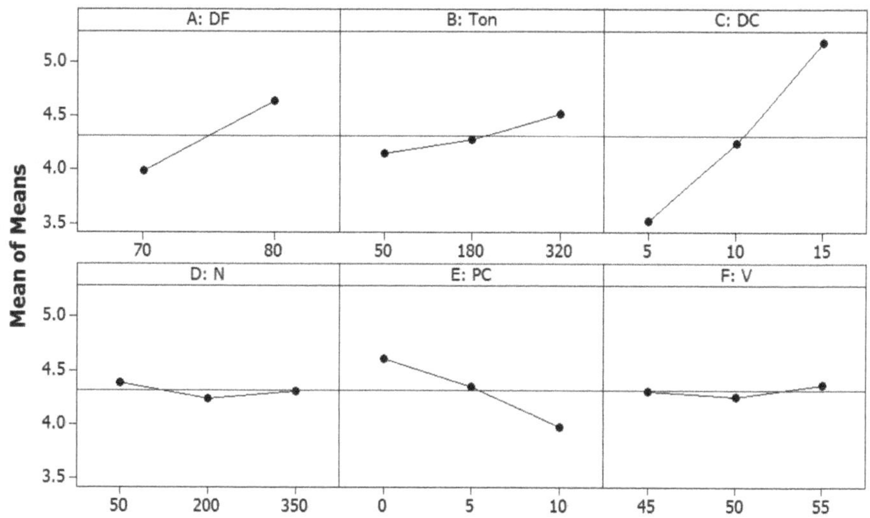

FIGURE 2.5 Effect of processing parameters on 3D surface area roughness, S_a (means).

and vaporization of the workpiece, thus increasing the S_a. However, the more surface roughness is observed with higher level of the duty factor, which leads to lower the frequency of solidification time that helps in the formation of deeper crater; as a result, the S_a increases.

It is revealed that the higher value of powder concentration decreases the roughness of the EDMed surface, and in the other way, it increases the *S/N* ratio of SR. It shows that an increase in the concentration of powder decreases the S_a. Mixing of powder to the dielectric medium minimizes the concentration of discharge at the localized spot that causes deeper crater. An increase in the powder concentration improves the uniform dispersion of discharge instead of localizing discharge that leads to swallow depth on the machined surface, thus decreasing the value of the S_a.

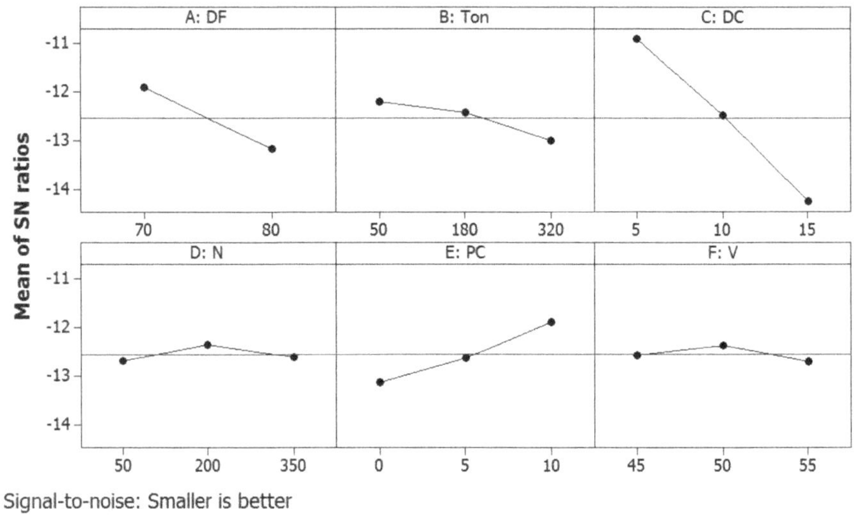

FIGURE 2.6 Effect of processing parameters on 3D surface area roughness, S_a (*S/N* ratios).

From Figures 2.5 and 2.6, it is also found that the mean of SR increased slightly, then reduced, and the increase in *S/N* ratio gradually results in higher values of tool rotational speed and gap voltage. Higher values of the tool rotational speed have less localized growth of sparks and enhance the centrifugal force in removing molten materials; thus, sparks are uniformly distributed on the machined surface and the molten metal at IEG is adequately removed with a minimal inclusion of unwanted particles resulting in the decrease in S_a.

ANOVA for S_a is analyzed, and the analysis results are given in Tables 2.5 and 2.6. It is observed that the tool rotational speed and gap voltage have less percentage of contribution (% CB), i.e., below 1%, and thus have a less effect on the 3D surface area

TABLE 2.5
ANOVA Results for 3D Surface Area Roughness, S_a (Means)

Source	DF	Seq SS	F	% CB	Remarks	Best Level
A: Duty factor	1	1.14206	312.24	16.03	***	$A_1 = 70$ %
B: Pulse on time	2	0.24436	33.4	3.43	*	$B_1 = 180$ μs
C: Discharge current	2	4.92874	673.77	69.17	***	$C_1 = 5$ A
D: Tool rotational speed	2	0.03822	5.22	0.54	×	$D_2 = 200$ rpm
E: Powder concentration	2	0.72843	99.58	10.22	**	$E_3 = 10$ g/L
F: Gap voltage	2	0.02142	2.93	0.30	×	$F_2 = 50$ V
Error	6	0.02195		0.31		
Total	17	7.12518				

***Most significant, **significant, *less significant, and × not significant.

TABLE 2.6
ANOVA Results for 3D Surface Area Roughness, S_a (S/N Ratios)

Source	DF	Seq SS	F	% CB	Remarks	Best Level
A: Duty factor	1	7.1456	114.2	14.67	***	$A_1 = 70\%$
B: Pulse on time	2	2.0509	16.39	4.21	*	$B_1 = 180\,\mu s$
C: Discharge current	2	33.7574	269.75	69.30	***	$C_1 = 5\,A$
D: Tool rotational speed	2	0.3898	3.12	0.80	×	$D_2 = 200\,rpm$
E: Powder concentration	2	4.6658	37.28	9.58	**	$E_3 = 10\,g/L$
F: Gap voltage	2	0.3275	2.62	0.67	×	$F_2 = 50\,V$
Error	6	0.3754		0.77		
Total	17	48.7125				

***Most significant, **significant, *less significant, and × not significant.

roughness, S_a. The discharge current is the most significant processing parameter for S_a followed by the duty factor. It is also revealed that powder concentration and pulse on time are also significant processing parameters. The lower value of the S_a can be achieved with the lower value of duty factor (A_1), pulse on time (B_1), and discharge current (C_1), and with the higher value of powder concentration (E_3). Thus, the optimal level of processing parameters is found as $A_1B_1C_1D_2E_3F_2$. The optimal level does not exist in the Taguchi orthogonal array. Thus, a confirmation test is conducted for the validity of the experiment. The three-experiment run is conducted on the optimal level, and their value should lie between the 95% confidence level ($\alpha = 0.05$).

2.4.3 Confirmation Test

The predicted grades at the optimum level values are calculated using Eq. (2.6).

$$(S_a)_{opt} = \sum_{1}^{n} \bar{X}_i - (n-1)\bar{M} = (\bar{A}_1 + \bar{B}_1 + \bar{C}_1 + \bar{E}_3) - 3\bar{M} \quad (2.6)$$

where $\bar{X}_i$ is the average mean of processing parameter X of level 'i'; $\bar{M}$ $\bar{X}_i$ is the average mean of processing parameter X of level 'i'; $\bar{M}$ is the overall mean of $S_a = 2.068\ \mu m$.

At 95% confidence level, the confidence interval (CI) is computed using Eq. (2.7) (Kansal, Singh, and Kumar 2007b).

$$CI = \pm\sqrt{\frac{F_{\alpha,(f_1,f_2)} \times V_{error}}{N_{eff}}} \quad (2.7)$$

where α is the significant level whose value is (1-confidence level); $F_{\alpha,(f_1,f_2)}$ is the variance ratio at α for f_1 and f_2 degrees of freedom (DOF); f_1 is the DOF of mean, whose value is 1; f_2 is the DOF of error = 10; N_{eff} is the number of participate factor test, whose value is given in Eq. (2.8); V_{error} is the pooled error variable of S_a; N is the total experiment number = 18.

$$N_{eff} = \frac{N}{1 + DOF} = \frac{18}{1+7} = 2.25 \tag{2.8}$$

From F-table of 95% confidence level, $F_{\alpha,(f_1,f_2)} = 4.9646$, and pooled variable error, $V_{error} = 0.0082$, from Table 2.5.

Now, the value of CI is ±0.008; therefore, the predicted mean of S_a at 95% confidence level is 2.068 ± 0.008 μm. Thus, the predicted S_a value is in the range of $2.060 < 2.068 < 2.076$ μm.

The experimental value of S_a at optimal processing parameters, $A_1B_1C_1D_2E_3F_2$, is 2.126 μm, which is very close to the predicted range of S_a. Thus, the experimental design is validated.

2.5 CONCLUSIONS AND SCOPE FOR FUTURE WORK

In this chapter, EDM with three-phase (deionized water–air–graphite powder) dielectric medium on the surface roughness was investigated and compared with single-phase (deionized water) and two-phase (deionized water–air) dielectric mediums. The important outcomes of the study are summarized as follows:

a. Use of the three-phase dielectric medium in EDM process achieves the least surface roughness parameters as compared to the single-phase and two-phase dielectric mediums.
b. In three-phase dielectric EDM, the values of 3D surface area roughness, S_a, reduce by 38.0% and 5.1%; the value of the maximum height of the surface, S_z, reduces by 36.5% and 3.4%, and the value of surface profile roughness, R_a, reduces by 15.5% and 5% as compared to EDM process with the single-phase and two-phase dielectric mediums, respectively.
c. Discharge current is the most significant factor on the 3D surface area roughness, S_a, followed by duty factor.
d. Increase in the powder concentration reduces the 3D surface area roughness, S_a, significantly.

Future work can be studied on the material removal mechanism and optimization of the processing parameters on the responses of EDM with three-phase dielectric medium. However, implementation of the EDM is still lacking in the industrial point of view due to the chances for inhaling and sprinkling the powder particles to the machine operator that may harm the operator. So, further study can be carried out to protect the operator hazards.

ACKNOWLEDGMENT

The authors have acknowledged IK Gujral Punjab Technical University, Punjab, and National Institute of Technology, Manipur, for their valuable support for conducting the experiment.

REFERENCES

Bai, X., T. Yang, and Q. Zhang. 2018. Experimental Study on the Electrical Discharge Machining with Three-Phase Flow Dielectric Medium. *The International Journal of Advanced Manufacturing Technology* 96, no. 5–8: 2003–2011. http://link.springer.com/10.1007/s00170-018-1747-x.

Bai, X., Q.H. Zhang, T.T. Li, and J.H. Zhang. 2012a. Powder Mixed Near Dry Electrical Discharge Machining. *Advanced Materials Research* 500: 253–258.

Bai, X., Q.H. Zhang, T.T. Li, and Y. Zhang. 2012b. Research on the Medium Breakdown Mechanism of Powder Mixed Near Dry Electrical Discharge Machining. *Chinese Journal of Mechanical Engineering* 48, no. 7: 186–192.

Batish, A., A. Bhattacharya, and N. Kumar. 2014. Powder Mixed Dielectric: An Approach for Improved Process Performance in EDM. *Particulate Science and Technology* 33: 37–41.

Bendell, A., J. Disney, and W.A. Pridmore. 1991. Taguchi Methods: Applications in World Industry. *Interfaces* 21, no. 2: 99–101.

Gao, Q., Q. Zhang, and J. Zhang. 2009. Experimental Study of Powder-Mixed near Dry Electrical Discharge Machining. *Chinese Journal of Mechanical Engineering* 45, no. 1: 169–175.

Gupta, K., N.K. Jain, and R.F. Laubscher. 2016. *Hybrid Machining Processes: Manufacturing and Surface Engineering*. SpringerBriefs in Applied Sciences and Technology. Cham: Springer.

Jeswani, M.L. 1981. Effect of the Addition of Graphite Powder to Kerosene Used as the Dielectric Fluid in Electrical Discharge Machining. *WEAR* 70, no. 2: 133–139. https://link.springer.com/article/10.1007/s00170-014-6433-z?no-access=true.

Kansal, H.K., S. Singh, and P. Kumar. 2007a. Technology and Research Developments in Powder Mixed Electric Discharge Machining (PMEDM). *Journal of Materials Processing Technology* 184, no. 1–3: 32–41.

Kansal, H.K., S. Singh, and P. Kumar. 2007b. Effect of Silicon Powder Mixed EDM on Machining Rate of AISI D2 Die Steel. *Journal of Manufacturing Processes* 9, no. 1: 13–22.

Khundrakpam, N.S., G.S. Brar, and D. Deepak. 2018. A Comparative Study on Machining Performance of Wet EDM, Near Dry EDM and Powder Mixed Near Dry EDM. *International Journal of Applied Engineering Research* 13, no. 11: 9378–9381.

Kumar, S., R. Singh, A. Batish, and T.P. Singh. 2015. Study the Surface Characteristics of Cryogenically Treated Tool- Electrodes in Powder Mixed Electric Discharge Machining Process. *Materials Science Forum* 808: 19–33.

Kumar, S., R. Singh, A. Batish, T.P. Singh, and R. Singh. 2017. Investigating Surface Properties of Cryogenically Treated Titanium Alloys in Powder Mixed Electric Discharge Machining. *Journal of the Brazilian Society of Mechanical Sciences and Engineering* 39, no. 7: 2635–2648.

Kunieda, M., M. Yoshida, and N. Taniguchi. 1997. Electrical Discharge Machining in Gas. *CIRP Annals - Manufacturing Technology* 46, no. 1: 143–146.

Lauwers, B., F. Klocke, and A. Klink. 2010. Advanced Manufacturing through the Implementation of Hybrid and Media Assisted Processes. *International Chemnitz Manufacturing Colloquium* 54: 205–220.

Lauwers, B., F. Klocke, A. Klink, A.E. Tekkaya, R. Neugebauer, and D. Mcintosh. 2014. Hybrid Processes in Manufacturing. CIRP Annals—Manufacturing Technology 63, no. 2: 561–583. http://dx.doi.org/10.1016/j.cirp.2014.05.003.

Marashi, H., D.M. Jafarlou, A.A.D. Sarhan, and M. Hamdi. 2016. State of the Art in Powder Mixed Dielectric for EDM Applications. *Precision Engineering* 46: 11–33.

Patel, S., D. Thesiya, and A. Rajurkar. 2018. Aluminium Powder Mixed Rotary Electric Discharge Machining (PMEDM) on Inconel 718. *Australian Journal of Mechanical Engineering* 16, no. 1: 21–30. https://doi.org/10.1080/14484846.2017.1294230.

Rajurkar, K.P., D. Zhu, J.A. McGeough, J. Kozak, and A. De Silva. 1999. New Developments in Electro-Chemical Machining. *CIRP Annals* 48, no. 2 (January 1): 567–579. https://www.sciencedirect.com/science/article/pii/S0007850607632351.

Road, G., and G. Park. 2016. Study the Effect of Black Layer on Electrode Wear Ratio in Powder Mixed Electric Discharge Machining of Titanium Alloys. *International Journal of Machining and Machinability of Materials* 18, no. 1/2: 18–35.

Singh, A.K., S. Kumar, and V.P. Singh. 2015. Effect of the Addition of Conductive Powder in Dielectric on the Surface Properties of Superalloy Super Co 605 by EDM Process. *The International Journal of Advanced Manufacturing Technology* 77, no. 1–4 (March 10): 99–106. http://link.springer.com/10.1007/s00170-014-6433-z.

Talla, G., S. Gangopadhyay, and C.K. Biswas. 2017. Influence of Graphite Powder Mixed EDM on the Surface Integrity Characteristics of Inconel 625. *Particulate Science and Technology* 35, no. 2: 219–226. https://www.tandfonline.com/doi/full/10.1080/02726351.2016.1150371.

Treppe, F., C. Hochmuth, R. Schneider, T. Junker, J. Schneider, and A. Stoll. 2011. Steigerung Der Ressourceneffizienz Durch Hybride Prozesse. *Nachhaltige Produktion*: 127–150.

Tzeng, Y.F., and C.Y. Lee. 2001. Effects of Powder Characteristics on Electrodischarge Machining Efficiency. *International Journal of Advanced Manufacturing Technology* 17, no. 8: 586–592.

Wong, Y.S., L.C. Lim, I. Rahuman, and W.M. Tee. 1998. Near-Mirror-Finish Phenomenon in EDM Using Powder-Mixed Dielectric. *Journal of Materials Processing Technology* 79: 30–40.

Yang, J.J., Z.J. Zuo, and W.X. Yu. 2010. Experimental Investigation on Air-Aided Water EDM. *Advanced Materials Research* 148–149: 471–474.

Zhao, W.S., Q.G. Meng, and Z.L. Wang. 2002. The Application of Research on Powder Mixed EDM in Rough Machining. *Journal of Materials Processing Technology* 129, no. 1–3: 30–33.

Zhu, Z., V.G. Dhokia, A. Nassehi, and S.T. Newman. 2013. A Review of Hybrid Manufacturing Processes—State of the Art and Future Perspectives. *International Journal of Computer Integrated Manufacturing* 26, no. 7: 596–615. https://doi.org/10.1080/0951192X.2012.749530.

3 Fabrication of Microchannels using Conventional and Hybrid Machining Processes

Atul Babbar, Vivek Jain, and Dheeraj Gupta
Thapar Institute of Engineering and Technology, Patiala

Ankit Sharma
Chitkara College of Applied Engineering,
Chitkara University, Patiala

CONTENTS

3.1 INTRODUCTION

Microchannels are miniatures grooving over the surface of the glass, polymers, and metallic substrate with the dimension less than 1 mm and greater than 1 μm. These microchannels are the only means for transporting liquid in biomedical and microfluidic devices. These are vitally important for the working of microfluidic devices [1–4]. Generally, glass and polymer substrates have been used for the various industrial and biomedical applications [5–13]. However, mechanical applications involve the use of the metallic substrate. The silicon-based substrates find their application in electronics industries. Microchannels possess certain advantages such as high surface-to-volume ratio and smaller volume. Microchannels have been used in numerous applications such as biotechnology, bioengineering, analysis of DNA and proteins, laboratory-on-chips, chemical reaction, mini-heat exchangers, and aerospace [14]. However, the level of precision required in fabricating these microchannels often creates different problems for the manufacturer. Furthermore, its utility in the form of cheaper and bulk need hinders its applicability owing to the lack of efficient technology and infrastructure. Until now, microchannels with different shapes and sizes have been fabricated using different techniques [1,2,6,15–41]. The common shapes are square (refers to Figure 3.1), rectangular, circular, U-shape (refers to Figure 3.2), V-shape (refers to Figure 3.3), and half-circle microchannels.

The circular shapes are fabricated inside the material, whereas the rest on the surface of a material. The important materials for microchannels are polymethyl methacrylate (PMMA) and polydimethyl siloxane (PDMS). PMMA has emerged as a promising solution for microfluidic devices. It has wider applications such as in drug delivery and implants owing to its excellent biocompatibility.

Miniaturization and rejection of heat at a faster pace lead to the usage of microchannels for cooling purpose in electronic and mechanical devices. The surface-to-volume

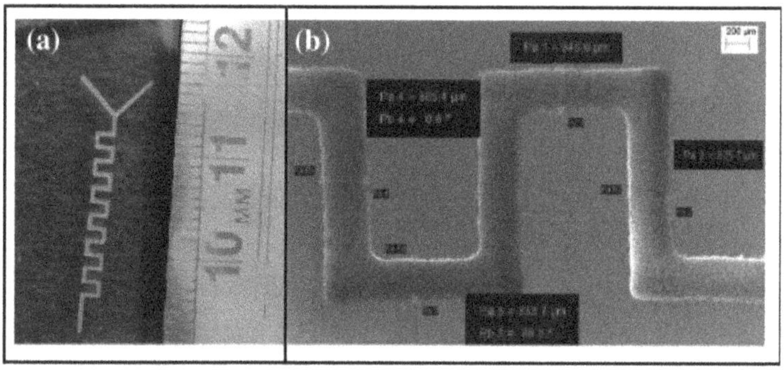

FIGURE 3.1 Microchannel with square profile (a) Square profile of serpentine microchannel. (b) FESEM image of micro-channel [26].

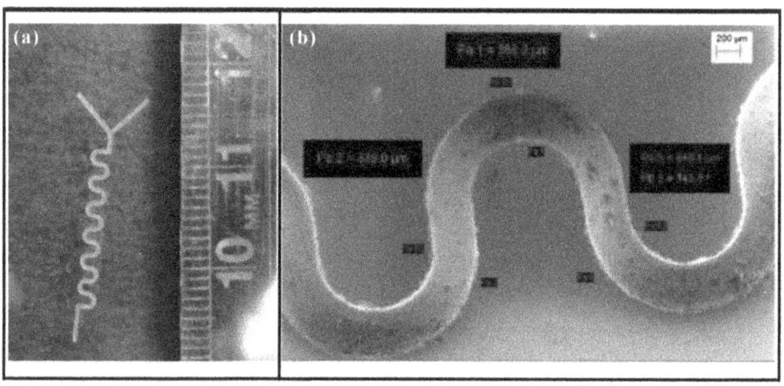

FIGURE 3.2 Microchannel with U-shape profile (a) Circular profile of serpentine microchannel. (b) FESEM image of micro-channel [26].

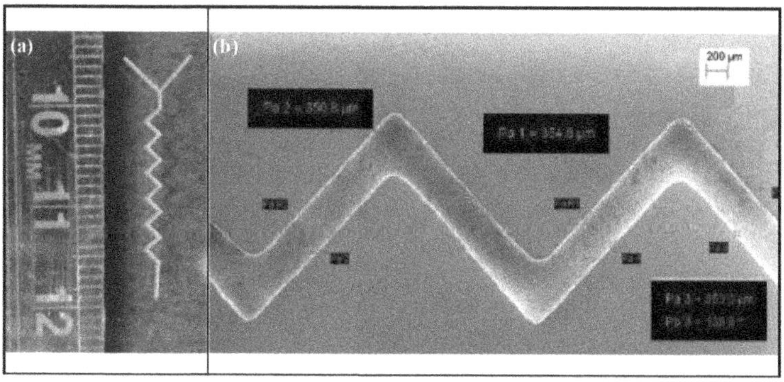

FIGURE 3.3 Microchannel with V-shape profile (a) Zig-zag profile of serpentine microchannel. (b) FESEM image of micro-channel [26].

ratio is increased by fabricating the microchannels over the metallic surface. The coolant of circuits is extremely important as they perform at a faster rate and higher temperature may lead to the disruption in the working of devices [42–44]. Apart from the silicon substrates, the quartz also possesses the features required to fabricate microchannels. Quartz is cheap, transparent, and chemically inert, and makes it suitable for microchannel-based devices. The method of fabrication of microchannels is extremely important since different technological and economical aspects are associated with it. The method used for fabricating the microchannels has been discussed in the subsequent sections.

3.2 CONVENTIONAL METHODS

The ability to implement a variety of materials for much smaller devices depends on the progress of micromachining and other production strategies. Different methods for different materials and applications have been studied worldwide. Different types of machining methods have been adopted for numerous applications [10,13,45–59]. This section covers some of the most commonly used fabrication processes.

3.2.1 Lithography

This is one of the most important manufacturing techniques for fabricating microchannels. The different types of topographies can be easily fabricated using technique known as photolithography, which assists in transferring the pattern into thin film. Martynova et al. [6] followed a simple lithographic approach to fabricate the round microchannels in PDMS with a diameter of 5–200 μm. Pal and Sato [7] used the single-stage lithography to fabricate various microstructure and microfluidic channel designs. Soft lithography was used to create microchannels in the PDMS. The PDMS curing takes place in the photoresist projected onto the surface of the silicon wafer. Subsequently, microchannels are formed by joining PDMS with the surface of glass. The fabrication of microchannels can also be performed on the polymeric material.

3.2.2 Wet and Dry Etching

Etching process has been used for the pattern replication of semiconductor devices by the removal of layer. Dry and wet etching types are the two most common methods of using etching for fabricating the microchannels. In wet etching, liquid-phase etchants are used to subtract the material from the wafer with the help of oxidizing agent. Diffusion and adsorption are the mechanisms that actively participate in the removal of the material from the surface. In dry etching, material removal takes place by means of either chemical, physical bombardment, or both. This method includes reactive ion, ion-beam induction, and sputtering.

3.2.3 Laser-Evolved Microchanneling

Since the complications of the process are worrying, laser-related microchannel manufacturing processes are the easiest because they have the potential to create any form of the microchannel. More flexibility, lesser time, and easy adaptability are its potential features. Usually, the process of removing the material depends on the heat. To get the best results, the process needs to be optimized. In particular, the laser production process requires no more than two steps, and there is no need to clean the room. There is no need to prepare the mask as is the case with photolithographic or engraving process.

The creation of channels with a smooth surface that is free of microcracks and edge chips has always been an interesting task because the laser mainly causes thermal stresses. Scientists have done a lot of research with selected laser sources to reduce the problem of thermally induced cracks. The material is removed from the surface of workpiece when laser beam impinges over the surface. Depending on wavelength, the photon energy changes. More energy is contained in photons with shorter wavelengths, whereas less energy is contained in photons with longer wavelengths. For this reason, a shorter wavelength is used more often than a longer laser beam for the transparent materials.

3.2.4 Powder Blasting

Erosion phenomena can be observed in many thermodynamic applications. Severe losses result from entities that come into contact with eroded particles that are present in the atmosphere or in a liquid flow. Some studies have been done on the erosion mechanisms of brittle materials such as glass and ceramics when hard and sharp corrosive materials collide. If the particle approaches the object at a certain speed and also has energy above the crack threshold, the local distortion of the material in the form of cracks is convinced, which ensures that some parts of the original material are removed. The relationship between the opening width of the mask and the depth of erosion is interrupted in order to produce microstructures of different depths in a microfluidic stage. The sandblasting technology consists of several steps, which creates a 3D microchannel in the glass. First, a photoresist, a photopolymer, or a metal mask is produced, which is essentially negative for the required microchannel in the glass.

TABLE 3.1
Different Conventional Techniques to Fabricate Microchannels.

Fabrication Method	Applications	Material	Key Points	Ref.
Injection molding	Electrochemistry, DNA elongation	Silicon metal polymers	Small-sized and low-strength material can be fabricated, weld lines	[60–62]
Embossing and imprinting	Biomechanical	Silicon metal polymers	High surface roughness, not suitable for bulk applications, high-temperature requirement	[6,32,34]
Lithography	Protein synthesis, blood and DNA analysis	Silicon metal polymers	Large lead time, skilled worker, clean room, high aspect ratio, complex topography	[48,63–65]
Wet–dry etching	Biomechanical	Metals and reactive materials	Poor precision, selective material removal, unparallel walls	[1,66–69]
Micro-mechanical cutting	Metallic microparts and heat sinks	Metals	Batch production not feasible, high accuracy, low surface roughness, cracks, burr formation	[59,70,71]

In general, elastomers and light-sensitive materials are used as masks due to their good resistance to dust jets. Due to its light-sensitive nature, the mask can be fabricated directly on a glass substrate using photolithography, making it easier to achieve several high-resolution functions. When the glass is finished with a photoresist, place it in front of the dust-cleaning nozzle from a distance. The part is kept constant to allow nozzle a motion in directions along a zigzag path. When it starts to work, it creates microchannels in the glass. Because of this limitation, an exact-size microchannel cannot be made. This leads to a higher surface roughness compared to other methods. This process is subject to several restrictions, which has the additional advantage that simple channels with a larger proportion can be created. Some key information regarding the fabrication of microchannels is provided in Table 3.1.

3.3 HYBRID MACHINING PROCESSES

3.3.1 Electrochemical Discharge Machining (ECDM)

ECDM is a hybrid machining process for the treatment of hard and brittle nonconductive materials in nature. It is a hybrid of electric discharge machining (EDM) and electrochemical machining (ECM) processes, which is generally used to machine hard and brittle nonconductive materials such as glass, ceramics, refractory bricks, quartz, and composite materials. It is a complicated physiochemical process in which the material is removed from the workpiece by the anodic dissolution of the material and electrical sparks that occur in the middle of the working surface of the electrode element and the electrode tool. The energy and electrical discharges generated by sparks lead to a series of microexplosions on the surface, which increase the temperature of the local point to a very high level, causing the material to melt and evaporate, leading to a partial chemical attack, and removing the material in microns.

3.3.2 Ultrasonic-Assisted Hybrid Machining

Rotary ultrasonic milling is one of the most economical and acceptable machining processes in the atmosphere of existing materials for machining advanced technical materials that improve hole accuracy, higher material removal rate (MRR), finer surface finish, and high tolerances. The configuration of the rotating ultrasonic milling essentially consists of an ultrasonic spindle system, a refrigerant supply system, and a data acquisition system as main features. The assembly consists of an ultrasonic wave, a power supply, a converter, and an electric motor. The power supply uses a high-performance sine wave generator, which converts the low-frequency (50–60 Hz) electrical signals into the high-frequency electrical signals (up to 20 kHz). A piezoelectric transducer receives this high energy and converts it into high-frequency linear mechanical vibrations. In general, the ultrasonic piezoelectric transducer is preferred because it offers better performance with highly efficient electromechanical changes (up to 96%), which also means less transducer cooling compared to a lower-efficiency (up to 20%) magnetostrictive transducer. requires. –35%). Since the magnetostrictive transducer converts electrical signals into mechanical vibrations in two stages, this leads to energy loss in the form of heat. The illustration of the micro-ultrasonic milling setup is shown in Figure 3.4. With the same output power, the

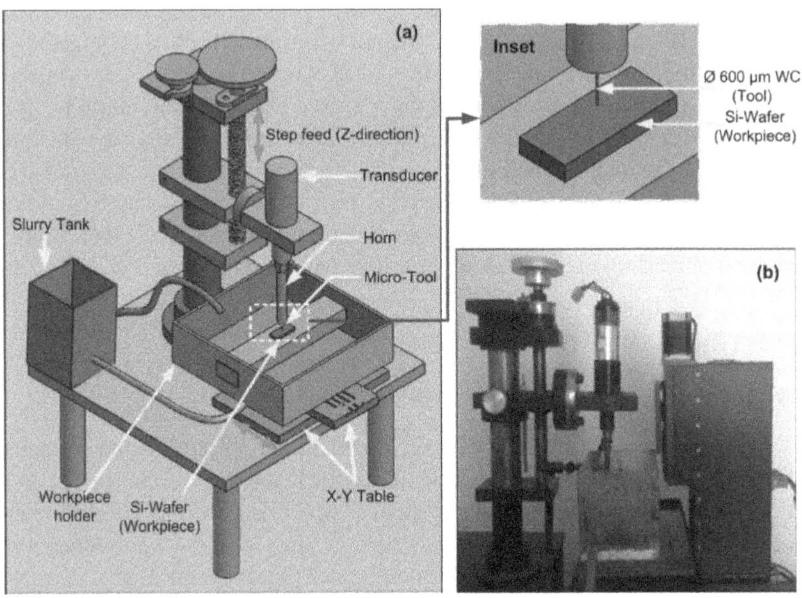

FIGURE 3.4 (a) Schematic representation of the micro-USM setup and (b) actual experimental setup [47].

piezoelectric converter consumes less energy, which also leads to low noise levels, so that the working conditions are more relaxed for the employee.

These ultrasonic vibrations are further amplified and then transferred to a tool coated with a diamond abrasive, whereby the tool vibrates at the ultrasonic frequency. It is possible to change the vibration amplitude by adjusting the control speed of the power supply output. The rotary movement of the cutting tool is achieved by starting an electric motor that is mounted on the top of the ultrasound axis. The engine speed can be controlled with the control unit button.

To investigate the complex mechanisms in material removal, many researchers examined the topography of a machined surface. The main reason for removing material from the work surface was considered a brittle crack. This main type of material removal was caused by a mixture of impact, abrasion, and extraction that occurred during the rotation and vibration of the tool. The overcut, which is the difference between the actual and nominal width of the microchannels, is shown in Figure 3.5.

The impact of the diamond grains was promoted as a better aspect when removing the material in the direction of the tooltip, while for the milled part, the grinding effect was mainly found near the walls. Hammering and abrasion caused deposits to build up, and subsequent mixing with the pressurized coolant was responsible for removing the milled surface walls. During fabrication of the microchannels, tool wear is another concern which needs to be addressed. Figure 3.6 shows the tool wear surface micrographs that highlight the deformation and wear in the tool during the fabrication of microchannels using an ultrasonic machining (USM) process.

As the top surface is compressed, the material undergoes lateral deformation. Consequentially, the diameter of the tool increases. Afterward, this deformed portion

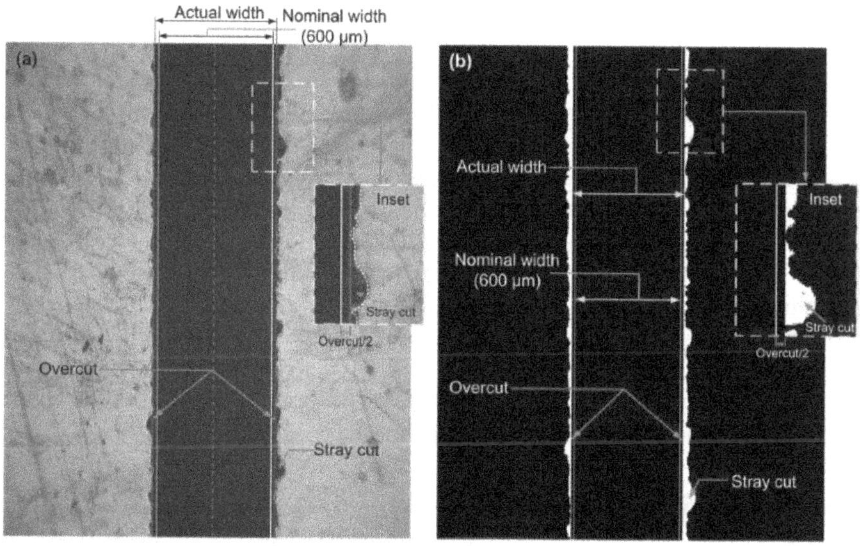

FIGURE 3.5 (a) Microchannel view and (b) stray-cut analysis of microchannel having black-and-white imaging [47].

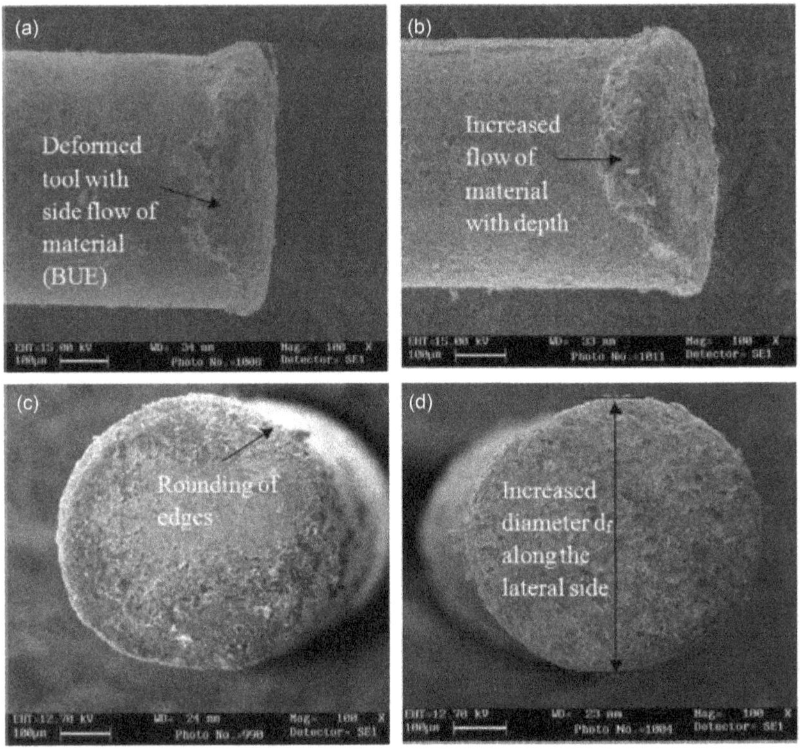

FIGURE 3.6 (a–b) Side view of the micro-USM tool and (c–d) top view after machining at different depth of cut [31].

becomes an integral part of the machining tool. This results in the inaccuracy and deviation from the actual dimensions of the microchannels. Subsequently, the material removal rate also decreases. It has been observed that very coarse and very fine size abrasives cause a lesser amount of form accuracy. Microchannels have also been fabricated on the thermoplastic substrate material, as shown in Figure 3.7.

The mold made of aluminum to manufacture the microreactor is represented in Figure 3.7a. The microreactor pattern replicated by ultrasonic embossing in PMMA medium is shown in Figure 3.7b, and the microreactor combined with injected colored water is shown in Figure 3.7c. Ultrasonic pressing takes place in two stages: In the first phase, the shape penetrates completely into the substrate amid hammering and rubbing. In the second phase, because the mold penetrates completely, the hammer stops and friction effectively improves the replication speed.

3.4 CASE STUDIES

Choi et al. [73] studied the chemical-assisted ultrasonic machining (CUSM), which overcame the drawbacks of USM such as low MRR and low surface quality. For this, in abrasive slurry, hydrofluoric (HF) acid was added in low concentration.

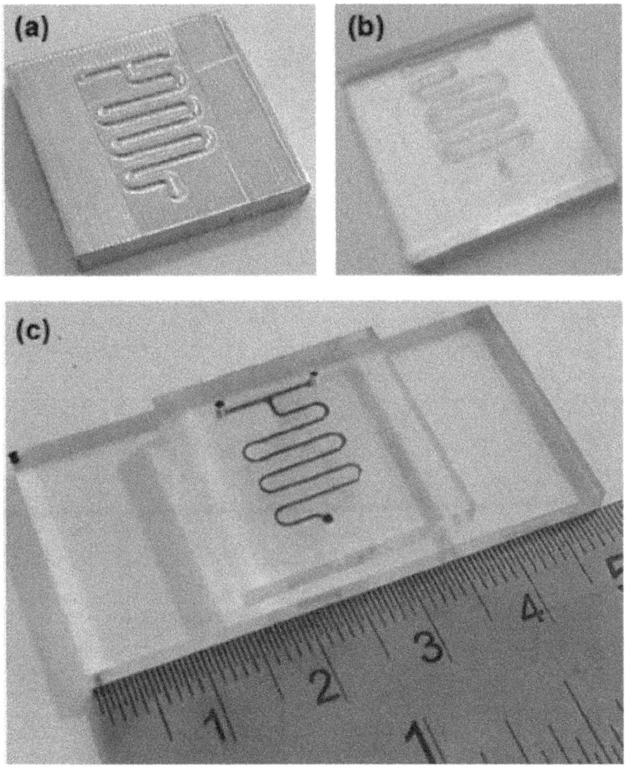

FIGURE 3.7 (a) Aluminum fold for microreactor chip, (b) pattern replicated on PMMA, and (c) chip with dyed water [72].

For obtaining the optimal conditions, the authors investigated the machining mechanisms and also carried out several experiments. From the different experiments and analyses, the authors concluded that there was an improvement in the surface roughness and the MRR by 40% at micro-drilling and 200% at macro-drilling. Also, the load of the machining was severely lowered and can be stably maintained. However, the size of the machined hole was enlarged by CUSM to a certain degree; therefore, it is suggested to utilize somewhat low concentration, i.e., below 5% HF.

Wang et al. [49] prepared the microchannels using UV lithography technique as coverslip as a substrate material. SU-8-2025 photoresist is used for lithography. Then, the coverslip is heated up to 95°C for 3 hours. The chemical stability was increased with the prebaking, which is at 95°C for 1 hour. Then, the samples were allowed to cool down under room temperature. The light source with UV wavelength is provided for 10 minutes. Subsequently, the photoresist was post-baked. The microturbine in the microchannel is covered with PDMS film, and finally, a sandwich-type structure is clipped off. The schematic representation of the microturbine with the corresponding surface electron microscopic images of the microchannels is shown in Figure 3.8.

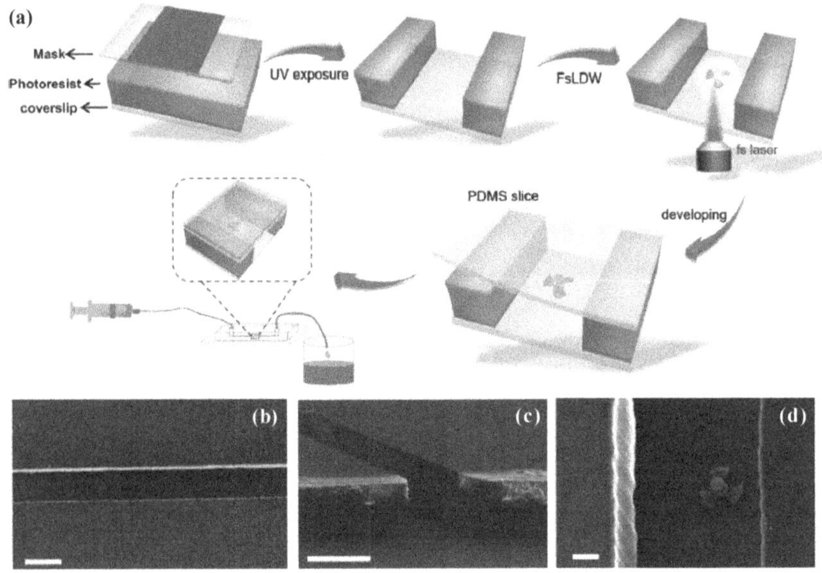

FIGURE 3.8 Schematic illustration of the microturbine with the corresponding surface electron microscopic images of the microchannels (a) Fabrication of micro turbine in microchannel. (b) and (c) SEM image of micro-channel at 100 microns. (d) SEM image of micro turbine [49].

Cao et al. [74] examined the potential of the ECDM machine, and produced and examined a glass microstructure of less than 100 μm. Research on this topic was necessary because the USM restrictions included severe tool wear and mechanical cracks on the surface and inside the structure. You can use this procedure to process non-conductive materials. Since the microstructures are less than 100 μm in size, they still pose a challenge with good surface quality due to the control of material removal rate, surface roughness, and machining resolution. As shown in Figure 3.9, mechanical cracks were observed at the edge of the channel in Section 3.2 because the tool was in contact with the work surface while serving. There was no existence of the cracks at a depth of 30 μm. It is said that the depth of the machining layer can be 30 μm or less.

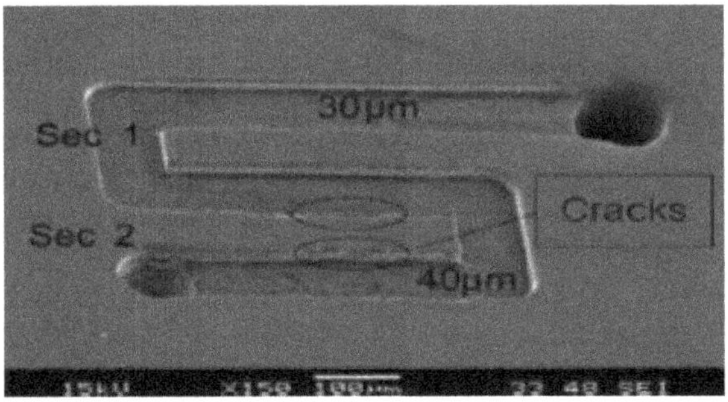

FIGURE 3.9 Mechanical cracks resulting from mechanical contact [74].

Prakash et al. [53] studied laser processing for the fabrication of microchannels on PMMA, which was carried out under water so that the heat-affected zone (HAZ), microcracking, and burr formation can be minimized. The authors used response surface methodology to conduct their experiments, and by using various parameters, the authors concluded that all the output parameters such as HAZ and burr formation depend on the processing parameters of the laser.

Atkin et al. [75] worked on PET to manufacture microfluidic devices, mainly using biochips for DNA diagnosis. Monitoring the demand for disposable diagnostic sensors in the healthcare sector leads to the development of inexpensive microfluidic devices. The authors worked on PET due to the limitations of polymer materials, and the desired properties for biochip surgery, such as the poor properties of electroosmotic flow and the high nonspecific binding, are also missing. With this technique, complex geometric shapes of approx. 10 μm can be created.

3.5 CONCLUSION

It has been observed that fabrication of the microchannels over polymer is a challenging task owing to its uncontrolled fracture. Moreover, metals do not provide necessary features such as optical transparency and nonreactive nature. However, the process becomes easier during the fabrication of the microchannels on the bone instead of polymers. Ceramics and semiconductor materials can be further studied for their potential application in the fabrication of microchannels. Fabrication of the microchannels using the conventional methods is not efficient as it hinders its rapid production. Automation and manufacturer skills are critically vital for the fabrication of microchannels on different grades of the materials. In lithography, there is always a need for the post-processing for high-resolution microchannels. During embossing, stamps regularly wear out results in the form of inaccuracy. Burr formation and heat-affected zone are of particular concern during the laser-based fabrication of microchannels. Furthermore, a lot of research is required to mitigate the effect of thermal damage during microchannel's fabrication. It has been observed that hybrid machining processes have great advantages towards the fabrication of microchannels in comparison with the conventional machining methods. There is a strong need to develop novel material on which microchannels can be easily fabricated with low production cost.

REFERENCES

1. T. Kikuchi, Y. Wachi, M. Sakairi and R.O. Suzuki, Aluminum bulk micromachining through an anodic oxide mask by electrochemical etching in an acetic acid/perchloric acid solution, *Microelectron. Eng.* 111 (2013), pp. 14–20.
2. S. Kumar and A. Dvivedi, On effect of tool rotation on performance of rotary tool micro-ultrasonic machining, *Mater. Manuf. Process.* 34 (2019), pp. 475–486.
3. J.K. Park, J.W. Yoon, M.C. Kang and S.H. Cho, Surface effects of hybrid vibration-assisted femtosecond laser system for micro-hole drilling of copper substrate, *Trans. Nonferrous Met. Soc. China* 22 (2012), pp. s801–s807.
4. K.K. Saxena, M. Bellotti, J. Qian, D. Reynaerts, B. Lauwers and X. Luo, *Overview of Hybrid Machining Processes*, Publisher Academic Press; Vol. 3, Elsevier Ltd, 2018.

5. A. Sharma, V. Jain and D. Gupta, Characterization of chipping and tool wear during drilling of Float glass using rotary ultrasonic machining, *Measurement* 128 (2018), pp. 254–263.
6. L. Martynova, L.E. Locascio, M. Gaitan, G.W. Kramer, R.G. Christensen and W.A. MacCrehan, Fabrication of plastic microfluid channels by imprinting methods, *Anal. Chem.* 69 (1997), pp. 4783–4789.
7. A. Sharma and V. Jain, Experimental investigation of cutting temperature during drilling of float glass specimen, in *IOP Conference Series: Materials Science and Engineering*, Shanghai, China, 2020.
8. A. Sharma, V. Jain and D. Gupta, Comparative analysis of chipping mechanics of float glass during rotary ultrasonic drilling and conventional drilling: For multi-shaped tools, *Mach. Sci. Technol.* 23 (2019), pp. 547–568.
9. A. Sharma, V. Jain and D. Gupta, Multi-shaped tool wear study during rotary ultrasonic drilling and conventional drilling for amorphous solid, *Proc. Inst. Mech. Eng. Part E J. Process Mech. Eng.* 233 (2019), pp. 551–560.
10. A. Sharma, A. Babbar, V. Jain and D. Gupta, Enhancement of surface roughness for brittle material during rotary ultrasonic machining, *MATEC Web of Conferences* 249 (2018), pp. 01006.
11. M. Kumar, A. Sharma and A.S. Shahi, A sensitization studies on the metallurgical and corrosion behavior of AISI 304 SS welds, in *Advances in Manufacturing Processes*, Lecture Notes in Mechanical Engineering, edited by K. S. Vijay Sekar, M. Gupta and A. Arockiarajan, 2018, pp. 257–265. Springer Nature Switzerland AG, Basel.
12. A. Sharma, V. Jain and D. Gupta, Tool wear analysis while creating blind holes on float glass using conventional drilling: A multi-shaped tools study, in *Advances in Manufacturing Processes*, Lecture Notes in Mechanical Engineering, edited by K. S. Vijay Sekar, M. Gupta and A. Arockiarajan, 2018, pp. 257–265.Springer Nature Switzerland AG, Basel.
13. M. Kumar, A. Babbar, A. Sharma and A.S. Shahi, Effect of post weld thermal aging (PWTA) sensitization on micro-hardness and corrosion behavior of AISI 304 weld joints, *J. Phys. Conf. Ser.* 1240 (2019), pp. 012078.
14. H.-W. Kang, S.J. Lee, I.K. Ko, C. Kengla, J.J. Yoo and A. Atala, A 3D bioprinting system to produce human-scale tissue constructs with structural integrity, *Nat. Biotechnol.* 34 (2016), pp. 312–319.
15. M. Pannipara, A.G. Al-Sehemi, A. Kalam and T.N. Mohammed Musthafa, Photophysics of dihydroquinazolinone derivatives: Experimental and theoretical studies, *J. Fluoresc.* 27 (2017), pp. 1161–1170.
16. N.D. Misal and M. Sadaiah, Multi-objective optimization of photochemical machining of Inconel 601 using grey relational analysis, *Mater. Today Proc.* 5 (2018), pp. 5591–5600.
17. C. Lee Ventola, Medical applications for 3D printing: Current and projected uses, *P T* 39 (2014), pp. 704–711.
18. T. Matsumura, Micromilling, *Compr. Mater. Process.* 11 (2014), pp. 151–177.
19. M. Yang, C. Li, Y. Zhang, D. Jia, X. Zhang, Y. Hou et al., Microscale bone grinding temperature by dynamic heat flux in nanoparticle jet mist cooling with different particle sizes, *Mater. Manuf. Process.* 33 (2018), pp. 58–68.
20. V. Jain, A.K. Sharma and P. Kumar, Recent developments and research issues in microultrasonic machining, *ISRN Mech. Eng.* 2011 (2011), pp. 1–15.
21. M. Rahman, A.B.M.A. Asad and Y.S. Wong, Introduction to advanced machining technologies, in *Comprehensive Materials Processing*, Editors-in-Chief **Saleem Hashmi**, Dublin, Ireland Elsevier, 2014, pp. 1–13.

22. J. Wang, J. Zhang, P. Feng and P. Guo, Damage formation and suppression in rotary ultrasonic machining of hard and brittle materials: A critical review, *Ceram. Int.* 44 (2018), pp. 1227–1239.
23. S. Shahane, N. Aluru, P. Ferreira, S.G. Kapoor and S.P. Vanka, Optimization of solidification in die casting using numerical simulations and machine learning, *J. Manuf. Process.* 51 (2020), pp. 130–141.
24. M. Yang, C. Li, Y. Zhang, Y. Wang, B. Li, D. Jia et al., Research on microscale skull grinding temperature field under different cooling conditions, *Appl. Therm. Eng.* 126 (2017), pp. 525–537.
25. C. Esposito Corcione, F. Gervaso, F. Scalera, S.K. Padmanabhan, M. Madaghiele, F. Montagna et al., Highly loaded hydroxyapatite microsphere/PLA porous scaffolds obtained by fused deposition modelling, *Ceram. Int.* 45 (2019), pp. 2803–2810.
26. S.S. Das and P.K. Patowari, Fabrication of serpentine micro-channels on glass by ultrasonic machining using developed micro-tool by wire-cut electric discharge machining, *Int. J. Adv. Manuf. Technol.* 95 (2018), pp. 3013–3028.
27. D.E. Brehl and T.A. Dow, Review of vibration-assisted machining, *Precis. Eng.* 32 (2008), pp. 153–172.
28. S.P. Leo Kumar, J. Jerald, S. Kumanan and R. Prabakaran, A review on current research aspects in tool-based micromachining processes, *Mater. Manuf. Process.* 29 (2014), pp. 1291–1337.
29. Z. Zhi-Jin, J.-Z. Zhu, B.J. Schaller, R. Gruber, H.A. Merten, T. Kruschat et al., Thermal analysis of grinding, *Proc. Inst. Mech. Eng. Part B J. Eng. Manuf.* 1 (2012), pp. 101–107.
30. M.P. Jahan, M. Rahman and Y.S. Wong, Micro-electrical discharge machining (micro-EDM), in *Comprehensive Materials Processing*, Editors-in-Chief Saleem Hashmi, Dublin, Ireland. Elsevier, 2014, pp. 333–371.
31. M.S. Cheema, A. Dvivedi and A.K. Sharma, Tool wear studies in fabrication of microchannels in ultrasonic micromachining, *Ultrasonics* 57 (2015), pp. 57–64.
32. R.M. McCormick, R.J. Nelson, M.G. Alonso-Amigo, D.J. Benvegnu and H.H. Hooper, Microchannel electrophoretic separations of DNA in injection-molded plastic substrates, *Anal. Chem.* 69 (1997), pp. 2626–2630.
33. A. Mogra, S.K. Verma and T. Thomas, Fabrication of square microchannel for experimental investigation of two phase flow using conventional machining process, *Perspect. Sci.* 8 (2016), pp. 231–233.
34. M.-C. Lin, J.-P. Yeh, S.-C. Chen, R.-D. Chien and C.-L. Hsu, Study on the replication accuracy of polymer hot embossed microchannels, *Int. Commun. Heat Mass Transf.* 42 (2013), pp. 55–61.
35. N. Celik, G. Pusat and E. Turgut, Application of Taguchi method and grey relational analysis on a turbulated heat exchanger, *Int. J. Therm. Sci.* 124 (2018), pp. 85–97.
36. M. Rahman, A.B.M.A. Asad, Y.S. Wong, M.P. Jahan and T. Masaki, Compound and hybrid micromachining processes, in *Comprehensive Materials Processing*, Editors-in-Chief Saleem Hashmi, Dublin, Ireland. Elsevier, 2014, pp. 89–112.
37. A. Senthil Kumar, M.R. Aravind Raghavendra, W.K. Neo and M. Rahman, Fast and fine tool servo for ultraprecision machining, in *Comprehensive Materials Processing*, Elsevier, 2014, pp. 61–88.
38. D. Jia, C. Li, Y. Zhang, M. Yang, X. Zhang, R. Li et al., Experimental evaluation of surface topographies of NMQL grinding ZrO_2 ceramics combining multiangle ultrasonic vibration, *Int. J. Adv. Manuf. Technol.* 100 (2019), pp. 457–473.
39. L. Processing, A. Yevtushenko and P. Grzes, *Encyclopedia of Thermal Stresses*, Springer, Dordrecht, 2014.
40. W. Jamróz, J. Szafraniec, M. Kurek and R. Jachowicz, 3D printing in pharmaceutical and medical applications, *Pharm. Res.* 35 (2018), pp. 1–22.

41. N.D. Sempertegui, A.A. Narkhede, V. Thomas and S.S. Rao, A combined compression molding, heating, and leaching process for fabrication of micro-porous poly(ε-caprolactone) scaffolds, *J. Biomater. Sci. Polym. Ed.* 29 (2018), pp. 1978–1993.
42. C.-C. Ho, G.-R. Tseng, Y.-J. Chang, J.-C. Hsu and C.-L. Kuo, External electric field-assisted laser percussion drilling for highly reflective metals, *Adv. Mech. Eng.* 5 (2013), pp. 156707.
43. B. Lauwers, F. Klocke, A. Klink, A.E. Tekkaya, R. Neugebauer and D. McIntosh, Hybrid processes in manufacturing, *CIRP Ann. Manuf. Technol.* 63 (2014), pp. 561–583.
44. H.Y. Zheng and Z.W. Jiang, Femtosecond laser micromachining of silicon with an external electric field, *J. Micromech. Microeng.* (2010), pp. 17001–17004.
45. R. Baraiya, A. Babbar, V. Jain and D. Gupta, In-situ simultaneous surface finishing using abrasive flow machining via novel fixture, *J. Manuf. Process.* 50 (2020), pp. 266–278.
46. A. Babbar, A. Kumar, V. Jain and D. Gupta, Enhancement of activated tungsten inert gas (A-TIG) welding using multi-component TiO_2-SiO_2-Al_2O_3 hybrid flux, *Measurement* 148 (2019), pp. 106912.
47. D. Sreehari and A.K. Sharma, On form accuracy and surface roughness in micro-ultrasonic machining of silicon microchannels, *Precis. Eng.* 53 (2018), pp. 300–309.
48. K. Ueno, H.B. Kim and N. Kitamura, Characteristic electrochemical responses of polymer microchannel-microelectrode chips, *Anal. Chem.* 75 (2003), pp. 2086–2091.
49. G. Wang, X.-L. Xu, Y. Yang, S.-X. Li, Y.-Z. Dai, H. Zhu et al., In microchannel driven micromotor by microfluid liquid as potential multi-functional devices towards Lab on a chip, *Optik (Stuttg).* 206 (2020), pp. 164312.
50. S. Kumar and A. Dvivedi, Effect of tool materials on performance of rotary tool micro-USM process during fabrication of microchannels, *J. Brazilian Soc. Mech. Sci. Eng.* 41 (2019), pp. 1–16.
51. D. Sreehari and A.K. Sharma, On thermal performance of serpentine silicon microchannels, *Int. J. Therm. Sci.* 146 (2019), pp. 106067.
52. A.R. Bahadorimehr, Y. Jumril and B.Y. Majlis, Low cost fabrication of microfluidic microchannels for Lab-On-a-Chip applications, in *2010 International Conference on Electronic Devices, Systems and Applications*, 2010, pp. 242–244.
53. S. Prakash, B. Acherjee, A.S. Kuar and S. Mitra, An experimental investigation on Nd:YAG laser microchanneling on polymethyl methacrylate submerged in water, *Proc. Inst. Mech. Eng. Part B J. Eng. Manuf.* 227 (2013), pp. 508–519.
54. A. Babbar, P. Singh and H.S. Farwaha, Parametric study of magnetic abrasive finishing of UNS C26000 flat brass plate, *Int. J. Adv. Mechatronics Robot.* 9 (2017), pp. 83–89.
55. A. Babbar, A. Sharma, V. Jain and A.K. Jain, Rotary ultrasonic milling of C/SiC composites fabricated using chemical vapor infiltration and needling technique, *Mater. Res. Express* 6 (2019), pp. 085607.
56. A. Babbar, V. Jain and D. Gupta, Neurosurgical bone grinding, in *Biomanufacturing*, Springer International Publishing, Editor Chander Prakash, LPU, India, Cham, 2019, pp. 137–155.
57. D. Singh, A. Babbar, V. Jain, D. Gupta, S. Saxena and V. Dwibedi, Synthesis, characterization, and bioactivity investigation of biomimetic biodegradable PLA scaffold fabricated by fused filament fabrication process, *J. Brazilian Soc. Mech. Sci. Eng.* 41 (2019), pp. 121.
58. A. Babbar, V. Jain and D. Gupta, Thermogenesis mitigation using ultrasonic actuation during bone grinding: a hybrid approach using CEM43°C and Arrhenius model, *J. Brazilian Soc. Mech. Sci. Eng.* 41 (2019), pp. 401.
59. Z. Wan, Y. Li, H. Tang, W. Deng and Y. Tang, Characteristics and mechanism of top burr formation in slotting microchannels using arrayed thin slotting cutters, *Precis. Eng.* 38 (2014), pp. 28–35.

60. M. Matteucci, T.L. Christiansen, S. Tanzi, P.F. Østergaard, S.T. Larsen and R. Taboryski, Fabrication and characterization of injection molded multi level nano and microfluidic systems, *Microelectron. Eng.* 111 (2013), pp. 294–298.
61. V. Piotter, K. Mueller, K. Plewa, R. Ruprecht and J. Hausselt, Performance and simulation of thermoplastic micro injection molding, *Microsyst. Technol.* 8 (2002), pp. 387–390.
62. G. Tosello, A. Gava, H.N. Hansen, G. Lucchetta and F. Marinello, Micro-nano integrated manufacturing metrology for the characterization of micro injection moulded parts, in *Proceedings of the 7th International Conference European Society for Precision Engineering and Nanotechnology, EUSPEN 2007*, 2007.
63. J.C. McDonald, D.C. Duffy, J.R. Anderson, D.T. Chiu, H. Wu, O.J.A. Schueller et al., Fabrication of microfluidic systems in poly(dimethylsiloxane), *Electrophoresis* 21 (2000), pp. 27–40.
64. P. Pal and K. Sato, Various shapes of silicon freestanding microfluidic channels and microstructures in one-step lithography, *J. Micromech. Microeng.* 19 (2009), pp. 055003.
65. E. Delamarche, A. Bernard, H. Schmid, A. Bietsch, B. Michel and H. Biebuyck, Microfluidic networks for chemical patterning of substrates: Design and application to bioassays, *J. Am. Chem. Soc.* 120 (1998), pp. 500–508.
66. O. Wolter, Micromachined silicon sensors for scanning force microscopy, *J. Vac. Sci. Technol. B Microelectron. Nanom. Struct.* 9 (1991), pp. 1353.
67. V. Maselli, R. Osellame, G. Cerullo, R. Ramponi, P. Laporta, L. Magagnin et al., Fabrication of long microchannels with circular cross section using astigmatically shaped femtosecond laser pulses and chemical etching, *Appl. Phys. Lett.* 88 (2006), pp. 191107.
68. J.H. Park, N.-E. Lee, J. Lee, J.S. Park and H.D. Park, Deep dry etching of borosilicate glass using SF6 and SF6/Ar inductively coupled plasmas, *Microelectron. Eng.* 82 (2005), pp. 119–128.
69. E. Belloy, S. Thurre, E. Walckiers, A. Sayah and M.A.. Gijs, The introduction of powder blasting for sensor and microsystem applications, *Sensors Actuators A Phys.* 84 (2000), pp. 330–337.
70. M. Pan, D. Zeng and Y. Tang, Feasibility investigations on multi-cutter milling process: A novel fabrication method for microreactors with multiple microchannels, *J. Power Sources* 192 (2009), pp. 562–572.
71. J.L. Liow, Mechanical micromachining: a sustainable micro-device manufacturing approach? *J. Clean. Prod.* 17 (2009), pp. 662–667.
72. F. Sucularli, M.A.S. Arikan and E. Yildirim, Investigation of process-affected zone in ultrasonic embossing of microchannels on thermoplastic substrates, *J. Manuf. Process.* 50 (2020), pp. 394–402.
73. J.P. Choi, B.H. Jeon and B.H. Kim, Chemical-assisted ultrasonic machining of glass, *J. Mater. Process. Technol.* 191 (2007), pp. 153–156.
74. X.D. Cao, B.H. Kim and C.N. Chu, Micro-structuring of glass with features less than 100μm by electrochemical discharge machining, *Precis. Eng.* 33 (2009), pp. 459–465.
75. M. Atkin, J.P. Hayes, N. Brack, K. Poetter, R. Cattrall and E.C. Harvey, Disposable biochip fabrication for DNA diagnostics, in *Proceedings of SPIE 4937, Biomedical Applications of Micro- and Nanoengineering*, 2002, pp. 125.

4 Modified Ultrasonic Machining Process

Rupinder Singh
NITTTR Chandigarh

Sudhir Kumar
Thapar Institute of Engineering and Technology, Patiala

CONTENTS

4.1 INTRODUCTION

Ultrasonic machining (USM) is one of the nonconventional machining techniques of great importance as it makes machining of hard materials and alloys easier. Titanium (Ti) is one of the toughest elements found on earth, whose machining has been a key issue for industries for many decades. In these investigations, a case study has been reported on USM of Ti15 alloy. The investigations were performed by six input parameters—namely, (a) tool materials (T.M), (b) slurry concentration (S.C), (c) slurry type (S.T), (d) slurry temperature (S.Temp), (e) power rating (P.R), and (f) slurry grit size (S.G.S) of USM—and three output responses—namely, (a) material removal rate (MRR), (b) tool wear rate (TWR), and (c) surface roughness (S.R). The output responses were furthered processed with statistical tool of analysis of variance (ANOVA) and response surface optimization using regression tool. It has been observed that for combined optimization of three outputs together, the best settings are as follows: T.M (Ti), S.C (25%), S.T (alumina), S.Temp (10°C), P.R (30%), and S.G.S (500 μm). From the contour plots of TWR taking P.R and S.Temp as input parameters, it has been observed that the maximum wear rate has been observed for 20 < S.Temp < 50 range and for P.R < 33. When MRR as output and S.Temp (*x*-axis) and P.R (*y*-axis) as input conditions were analyzed by contour plot, it was observed that S.Temp in the range of 22°C–55°C and P.R > 70% have provided the maximum MRR. When S.R was taken as output condition for the same input parameters for contour plotting, it has been observed that for the range of 15 < S.Temp < 35 and 40 < P.R < 70%, the maximum contribution towards S.R was observed. Finally, hybridization of USM with other possible nonconventional technique (which may give fruitful results and represent a solution for accurate and precise machining of hard and tough materials) has been presented as a novel method.

4.2 PREVIOUSLY REPORTED STUDIES ON USM

Two broad categories of machining are available today in industry: One is conventional and other is nonconventional. Conventional part deals with the standard machining procedures with conventional machines such as lathe machine, drilling machine, shaper machine, and vertical milling machine, whereas nonconventional machines also perform the same activity but with different techniques such as electron discharge machining (EDM), abrasive jet machining (AJM), laser machining, and USM [1]. USM is one of the nonconventional machining processes, which is now common in industry to machine very hard materials such as tungsten, carbide materials, high-speed steel, and titanium. This process is generally used in automotive industries for intricate machining such as parts for aeronautical industries, space industries, and robotics [2–8]. Intricate shapes and size of holes can be formed with USM, whereas some constraints are also there like diameter-to-depth ratio for machining process [4,8]. Since 1927, USM has been emerging as a promising technique of nonconventional machining [9,10]. Also, the American researchers were the first to get patent on this technology in 1945 [3,11–12]. Various acronyms have been set for USM depending on the machining and input parameters used by USM, such as abrasive-assisted machining, ultrasonic-assisted cutting, and abrasive slurry-assisted drilling [13]. But in 1950, it basically emerged as a grinding process, which has been assisted by ultrasonic vibrations [4,11,14–15]. Machining toughest and hardest material was one of the challenges in front of industrial community; thus, USM emerged as a solution for the same [1]. Basic USM process starts with low electrical energy signals of poor frequency, which are then converted into high-frequency energy and fed to transducer that ultimately generates vibration [1,13–18]. Horn of machine focuses the vibration towards tool of machine, whereas the tool with vibration moves in reciprocation with high-frequency vibration, which is generally greater than 20 kHz and therefore known as ultrasonic vibrations [19–21]. Design of tool and horn is critical for the output properties as it is related to the accuracy of machining. Some P.R instruments may be used to control the input power [22]. Various abrasive slurries may be used, such as silicon carbide, oil-suspended abrasive slurry, boron carbide, and alumina-suspended slurry for material removal action in contact with the tool and the workpiece [13]. Tool vibration with slurry on the workpiece creates erosion action, and the material starts eroding from the surface, thus causing finishing action [23]. One more variant of USM—identical to USM—that has been developed is rotary ultrasonic machining (RUM), which is beneficial in terms of low power requirement and better output performances like better surface finish. Recently, RUM has been used for the investigation of machining characteristics for the ceramic matrix of materials and has provided satisfactory performance in terms of MRR [24–27]. EDM assisted with USM has been used for microdrilling of holes in titanium alloys, which was previously impossible with conventional processes—this drawback has given a new idea of combining various nonconventional machining techniques for the best output characteristics [17,19,21,28–29]. Various researchers have investigated turning operation with USM and have obtained better results with an effective reduction in cycle time of machining process [8,29,32–33]. Tool of required shape, and generally workpiece of required shape, is made up of silver brazing operation and is directed towards workpiece continuously assisted with high-frequency vibrations, and slurry in between the workpiece and the tool generates the erosive action and performs machining [1,3,6,13, 30–35]. Nonconventional operation of die sinking analogous to USM operation has been previously used by researchers to generate hole of three-dimensional (3D) cavities in titanium alloy [36–39]. MRR in case of USM is very low in comparison with other nonconventional techniques, but its capability of generating multiple holes with single pass on workpiece makes it competitive economically [1]. Time reduction for machining process is one of the basic advantages of using USM. Previous researchers have found 19 hours of reduction [40] in preparing graphite electrode for EDM using USM than the conventional copy milling process [41–45]. A number of researchers are now using CNC-assisted USM for intricate machining operation with high accuracy, but the cost, availability, and capacity to control the machine are some of the issues for CNC-assisted USM technique [13,23].

From the last 200 years, titanium (Ti) is one of the elements present in periodic table with atomic number 22 and has been used in high-strength application due to its excellent properties such as high toughness and strength. The applications of titanium were considered specifically for aeronautical parts or space parts due to high performance of metal. Processing of Ti was very less before 1950, but with the invention of USM and other nonconventional processes, machining capabilities of Ti have been explored by various researchers. It has been found that USM machining operation is capable of processing Ti element and its alloys, which furthered the applications of Ti in various other industries such as jet engine manufacturing and airframe components [46]. Alloys of Ti are very hard and difficult to be machined with tool as they generate high temperature when continuously machined with direct tool, thus affecting the tool life [47–48]. USM gives the solution for the problem and removes the direct contact of tool with the workpiece, and in place of direct contact, abrasive slurry is used which is continuously fed through pump; thus, the temperature of slurry is also low, resulting in better surface properties. Similar results with better MRR have been obtained by previous researchers using EDM for Ti alloys [49–52]. A combination of USM and EDM represents such a technology, which can eradicate the problematic areas from USM and can provide with excellent results [53]. In conventional drilling, unsupported length of drill bit creates a problem in hole, and the deflection of tool inside hole while processing creates poor surface finish, which is eradicated in case of USM where no unsupported length is present, and though, easy flow of chip takes place through the hole, thus generating proper machined surface [54–62].

From the literature survey, it has been found that many researchers have done an investigation on USM for different applications such as drilling and machining. The combination of USM with various other nonconventional techniques, especially EDM, has been explored previously, but hitherto very few works have been reported on the investigation of machining capabilities of titanium metal (Titan15) with different tool materials (T.M), S.C, S.Temp, slurry grit size, P.R, and S.T as input parameters. The effect of these input parameters have been explored in this study to make remarks on machining capabilities of Ti15 in comparison with other T.M. Further, ANOVA has been used for the optimization of output results for the standard input conditions.

4.3 MATERIALS AND METHODS FOR FIRST CASE STUDY ON MACHINING OF TI WORKPIECE ON USM

In this study, six input parameters have been taken into consideration, namely, (a) T.M, (b) S.C, (c) S.T, (d) S.Temp, (e) P.R, and (f) S.G.S. The three output responses have been measured, namely, (a) MRR, (b) TWR, and (c) S.R. Taguchi mixed-model L18 orthogonal array (OA) has been used, in which three levels of each input parameter have been taken to optimize the outputs using MiniTab 17 software package. Table 4.1 shows the levels of input parameters used in this study, and Table 4.2 gives the design of experimentation (DOE) used for investigation. Figure 4.1 shows the steps involved in this investigation.

TABLE 4.1
Levels of Input Parameters using Taguchi L18 OA

S.No.	1	2	3	4	5	6
Input	T.M's (stainless steel, high speed steel (HSS), high carbon steel (HCS), carbide, diamond, titanium)	S.C (%)	S.T	S.Temp (°C)	P.R (%)	S.G.S (micron)
Levels	Three levels of each	(a) 15 (b) 20 (c) 25	(a) Boron carbide (b) Silicon carbide (c) Alumina	(a) 10 (b) 27 (c) 60	(a) 30 (b) 60 (c) 90	(a) 220 (b) 320 (c) 500

TABLE 4.2
Taguchi L18 OA-Based DOE

S.No.	T.M	S.C	S.T	S.Temp	P.R (of 500W)	S.G.S
1	S.S	15%	Boron carbide	10°C	30%	220
2	S.S	20%	Silicon carbide	27°C	60%	320
3	S.S	25%	Alumina	60°C	90%	500
4	H.S.S	15%	Boron carbide	27°C	60%	500
5	H.S.S	20%	Silicon carbide	60°C	90%	220
6	H.S.S	25%	Alumina	10°C	30%	320
7	H.C.S	15%	Silicon carbide	10°C	90%	320
8	H.C.S	20%	Alumina	27°C	30%	500
9	H.C.S	25%	Boron carbide	60°C	60%	220
10	Carbide	15%	Alumina	60°C	60%	320
11	Carbide	20%	Boron carbide	10°C	90%	500
12	Carbide	25%	Silicon carbide	27°C	30%	220
13	Diamond	15%	Silicon carbide	60°C	30%	500
14	Diamond	20%	Alumina	10°C	60%	220
15	Diamond	25%	Boron carbide	27°C	90%	320
16	Titanium	15%	Alumina	27°C	90%	220
17	Titanium	20%	Boron carbide	60°C	30%	320
18	Titanium	25%	Silicon carbide	10°C	60%	500

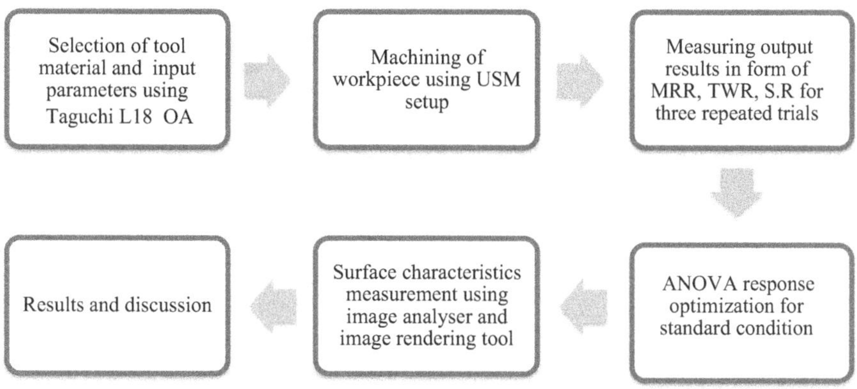

FIGURE 4.1 Steps involved in case study 1 of USM machining for Ti15.

4.4 EXPERIMENTATION

4.4.1 Ultrasonic Machining

USM with available setup with schematic of USM (see Figure 4.2a and b) of Ti15-based workpiece (dimensions as provided by supplier; see Figure 4.3) was performed with different sets of condition as explained in DOE (see Table 4.2).

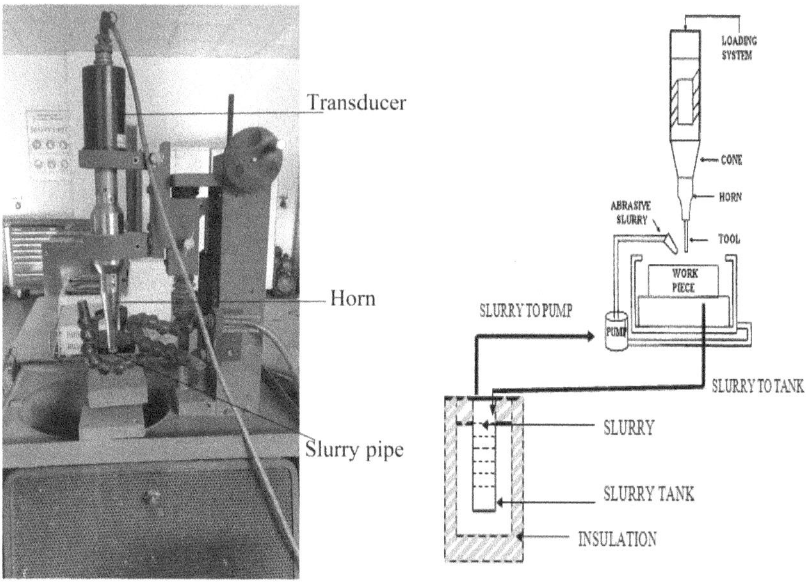

(a) (b)

FIGURE 4.2 (a) USM setup. (b) Schematic of USM.

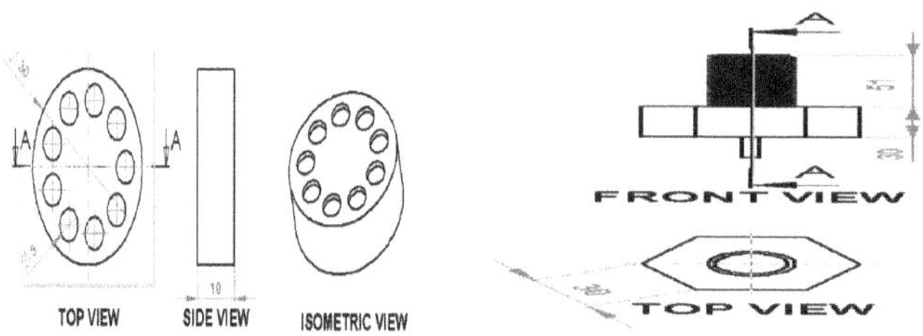

(a) (b)

FIGURE 4.3 (a) Workpiece design (material Ti15). (b) Design of tool.

4.4.2 MRR, TWR, and S.R Evaluation

After machining with USM, MRR was calculated based on the difference between weights of unmachined and machined workpieces. Similarly, TWR and S.R of the workpiece were evaluated using the S.R instrument (Talysurf) available in laboratory. Three sets of experimentation have been conducted for each experimental setup, and the average values have been used for further ANOVA.

4.4.3 Statistical Analysis and Surface Characterization

The obtained results have been analyzed using ANOVA optimization tool, and the photomicrographs taken from image analyzer have been processed for further surface characterization to predict the behavior of the output parameters in respect of the input parameters.

4.5 RESULTS AND DISCUSSION

4.5.1 USM Results

The workpiece of Ti15 (commercially pure) has been machined using six different tools (see Table 4.2) with selected DOE, and MRR, TWR, and S.R have been evaluated as discussed in 'Experimentation' section. The different results obtained for USM are given in Table 4.3. There is a large difference between USM-machined surface and conventionally machined surface, as shown in Figure 4.4. From Table 4.3, it was predicted that the properties were optimized using ANOVA tool of statistical analysis with Minitab 17 software package tool. From ANOVA results (see Table 4.4), it has been found that the residual error was 2.21% of the total error, which indicated that the model prediction was very accurate as the percentage residual error was less than 5% of the total error. It has been found that T.M showed major contributions towards MRR with highest percentage (52.93%) and S.G.S had the least role (1.27%)

TABLE 4.3
USM Machined Results for MRR, TWR, and S.R

S.No.	MRR	TWR	S.R
1	2.063	5.157	0.590
2	3.637	8.747	0.353
3	4.753	9.513	0.463
4	0.840	1.620	0.803
5	1.267	2.063	0.997
6	0.377	0.570	0.630
7	1.443	5.263	0.537
8	0.080	0.047	0.380
9	1.783	6.033	0.673
10	0.177	0.373	0.863
11	1.297	6.313	0.540
12	0.290	1.653	1.310
13	0.403	0.403	0.800
14	0.950	0.317	0.730
15	3.043	7.393	0.880
16	4.030	6.433	0.947
17	2.933	1.560	0.990
18	4.807	5.357	0.473

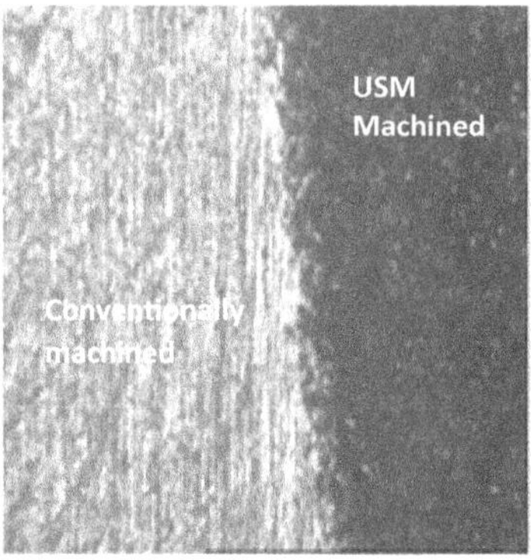

FIGURE 4.4 Comparative view of USM and conventionally machined surface.

TABLE 4.4
ANOVA for MRR for Larger Is the Better Case

Source	DF	Seq SS	Adj SS	Adj MS	F	P	% contribution
T.M	5	973.54	973.54	194.71	9.55	0.098	52.93
S.C	2	60.30	60.30	30.15	1.48	0.403	3.28
S.T	2	223.79	223.79	111.89	5.49	0.154	12.17
S.Temp	2	26.04	26.04	13.02	0.64	0.610	1.42
P.R	2	491.65	491.65	245.82	12.05	0.077	26.73
S.G.S	2	23.34	23.34	11.67	0.57	0.636	1.27
Residual error	2	40.79	40.79	20.39			
Total	17	1839.44					

to play towards MRR for the selected DOE. The rank table as given in Table 4.5 shows that T.M has been given first rank in output as it contributed most; similarly, S.G.S has been given sixth rank due to its least contribution. From the main-effect plot (see Figure 4.5), it has been observed that T.M of level 6, S.C of level 3, S.T of level 1, S.Temp of level 1, P.R of level 3, and S.G.S of level 1 are the optimized conditions for larger is the better case. But the optimized conditions were different from the selected list of DOE; therefore, confirmatory experiments were necessary to find the actual results on the optimized conditions. From the main-effect plots of different properties (see Figures 4.5–4.7), it was observed that for all properties, there were different sets of input conditions, which means there was further need of multioptimization as for different properties, we needed single setting so that all output parameters may be controlled with single processing condition. Therefore, we further analyzed the obtained result with regression technique of response surface optimizer, and it has been found that for the three outputs together, the best settings were found to be T.M of level 6, S.C of level 3, S.T of level 3, S.Temp of level 1, P.R of level 1, and S.G.S of level 3 and they altogether play a vital role in multioptimization of properties (see Figure 4.8). It has been observed that the condition for

TABLE 4.5
Rank Table for MRR for Larger is the Better Case

Level	T.M	S.C	S.T	S.Temp	P.R	S.G.S
1	10.3485	–0.4789	5.1787	2.7410	–5.5699	2.3796
2	–2.6473	0.4146	1.9088	–0.2029	2.1431	1.6463
3	–4.5754	3.7726	–3.3792	1.1702	7.1350	–0.3177
4	–7.8508					
5	0.4449					
6	11.6967					
Delta	19.5475	4.2515	8.5580	2.9439	12.7049	2.6973
Rank	1	4	3	5	2	6

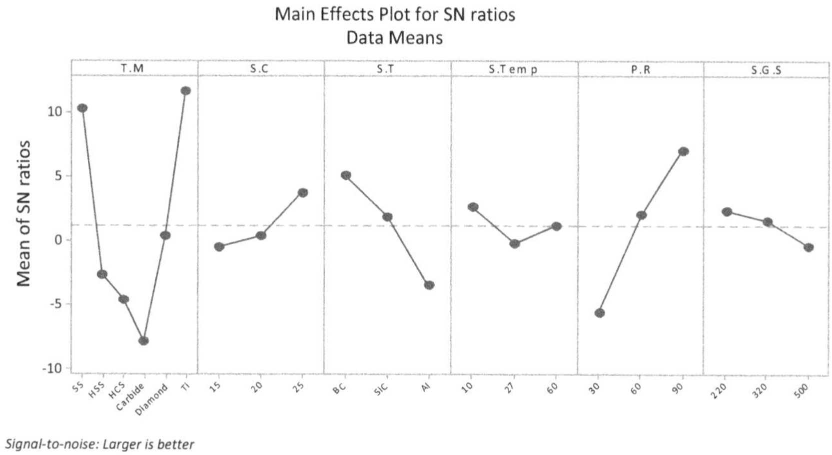

FIGURE 4.5 Main-effect plot for Signal to Noise (SN) ratio of MRR.

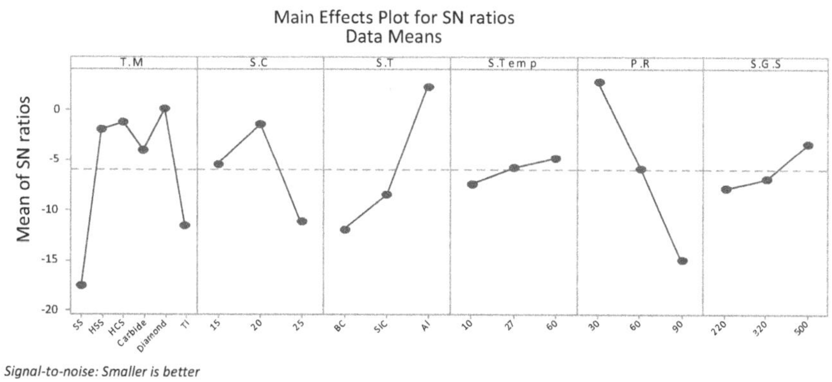

FIGURE 4.6 Main-effect plot for TWR.

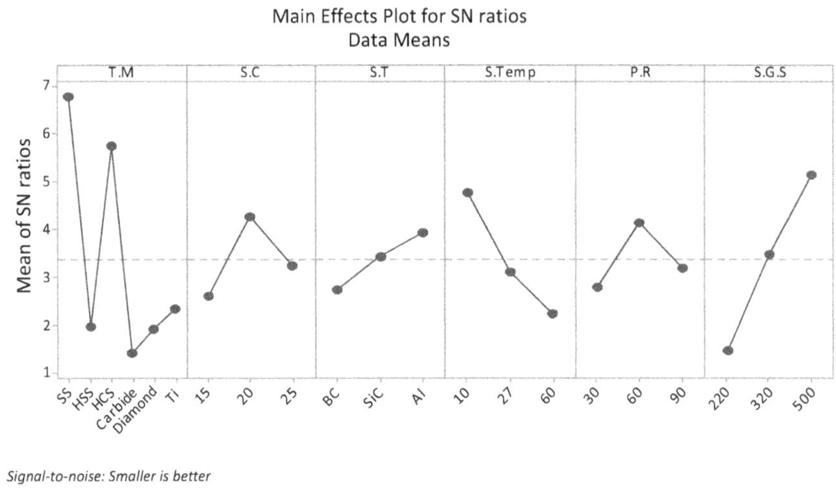

FIGURE 4.7 Main-effect plot for S.R.

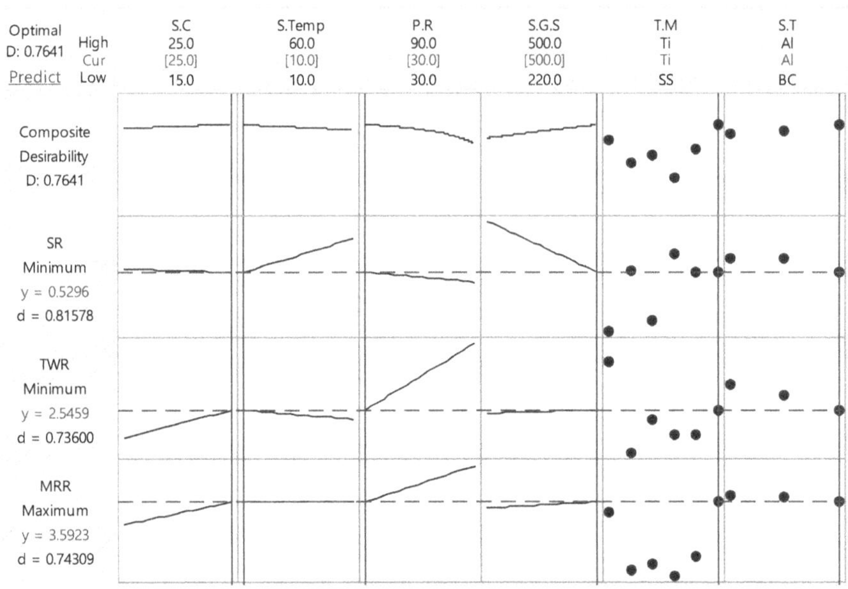

FIGURE 4.8 Response surface optimization for MRR, TWR, and S.R.

multioptimization was out of DOE, but the maximum input parameters (T.M, S.C, S.Temp, and S.G.S) resulted in greater contribution towards the output parameters as suggested by multi-optimization that were present in experimental condition 18. Thus, we may choose the experimental condition 18 as a standard condition of processing for Ti15 workpiece.

Further multioptimized data was used for plotting of contour (see Figure 4.9) taking MRR as output and S.Temp (x-axis) and P.R (y-axis) as input conditions while keeping other input parameters constant; it was observed that S.Temp in range of 22°C–55°C and P.R greater than 70 have provided nominal MRR. Similarly, when TWR has been taken as output property for the same setting of contour plot taking P.R and S.Temp as input parameters, it was observed that maximum wear rate has been obtained for 20 < S.Temp < 50 range and for P.R that was found to be less than 33 for the nominal results (see Figure 4.10). When S.R was taken as output property for the same input parameters for contour plotting, it was observed that the range of 15 < S.Temp < 35 and 40 < P.R < 70 have the maximum contribution towards S.R (Figure 4.11).

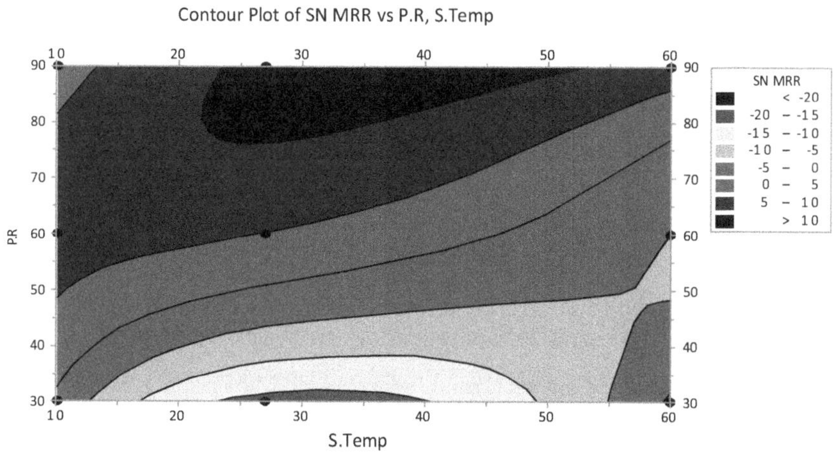

FIGURE 4.9 Contour plot for MRR taking P.R and S.Temp as input conditions.

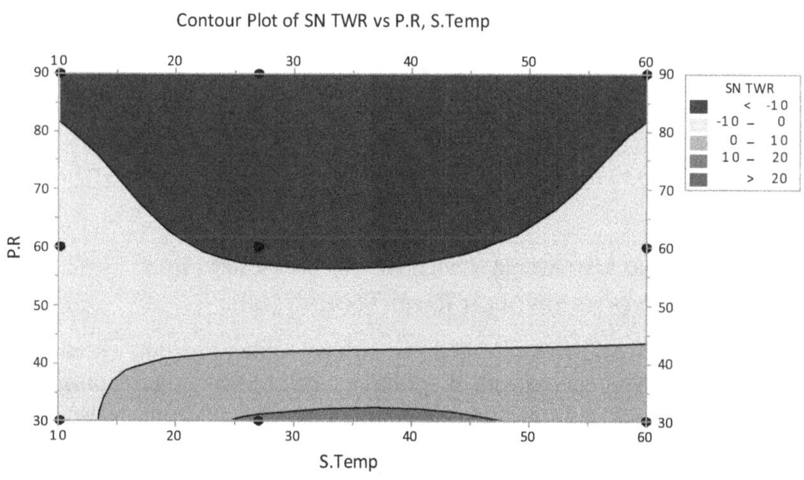

FIGURE 4.10 Contour plot for TWR taking P.R and S.Temp as input conditions.

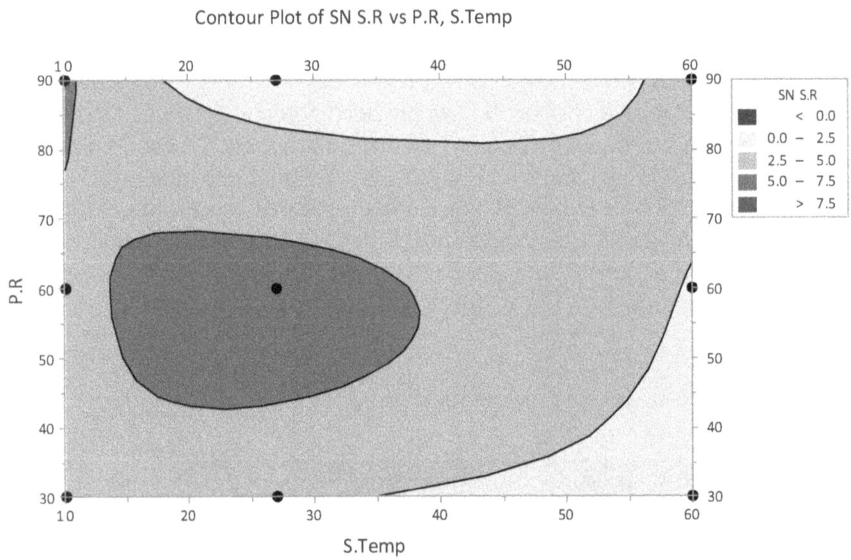

FIGURE 4.11 Contour plot for S.R taking P.R and S.Temp as input conditions.

4.5.2 Surface Characterization of Processed Workpiece

The 'Image Analyzer' (Metatech (Pune) India; Model: Metagraph-1; Magnification: 100×) has been used for taking photomicrographs of the processed workpiece (as per Table 4.2), and the obtained photomicrographs were further processed using image processing tool to render the surface characterization of the processed surface (see Figure 4.12).

4.6 POSSIBLE HYBRIDIZATION OF USM

Sections 4.6.1–4.6.5 represent some of the possible hybridization of USM with other nonconventional machining processes as per Indian Patent Application No: 202011001505 dated January 13, 2020, TEMP/E-1/1644/2020-DEL [63].

4.6.1 Combining Ultrasonic Drilling and Broaching as a Hybrid Process through Rapid Tooling [63]

The first possible hybridization of USM is with broaching process, where the rapid tool of USM may be prepared with 3D printing machine using thermoplastic matrix as base of tool and foreign particles of abrasive as reinforcement to thermoplastic matrix, which may provide broaching action while in use. Figure 4.13 shows the possible hybridized model of USM-based broaching. The possible details of the suggested hybridized process are provided in Figure 4.13. The total length of the tool is divided into four sections. From the tool tip (6) side, the first section is called rough

<table>
<tr><td></td><td>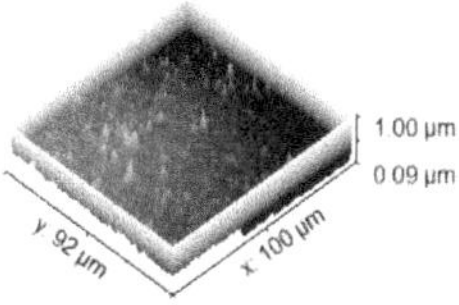
</td><td>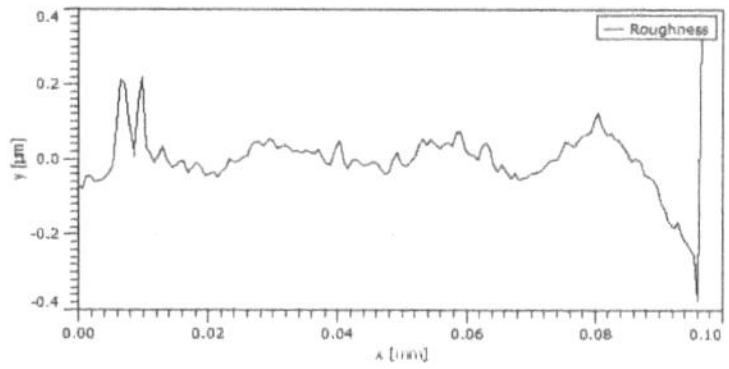
</td></tr>
<tr><td>Figure 4.12(a) [Exp. No.1]</td><td>Rendered image [Exp. No.1]</td><td>Ra 0.590 µm [Exp. No.1]</td></tr>
<tr><td></td><td>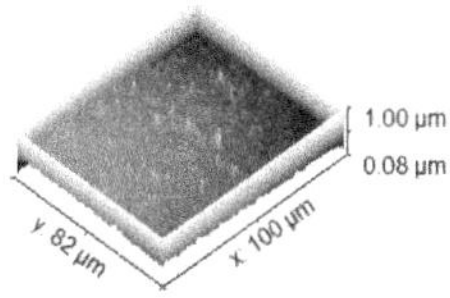
</td><td>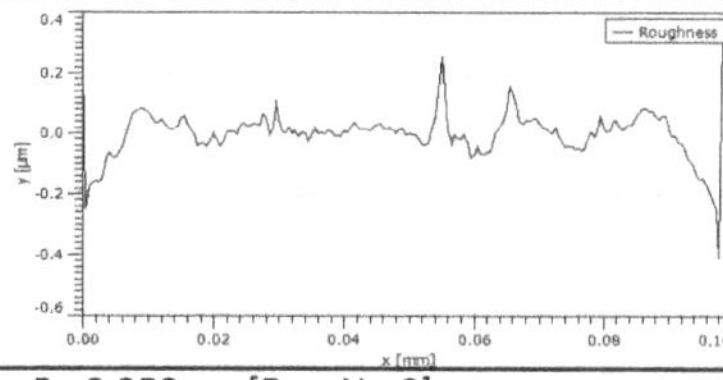
</td></tr>
<tr><td>Figure 4.12(b) [Exp. No.2]</td><td>Rendered image [Exp. No.2]</td><td>Ra 0.353 µm[Exp. No.2]</td></tr>
<tr><td></td><td>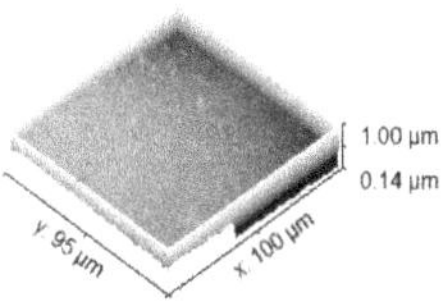
</td><td>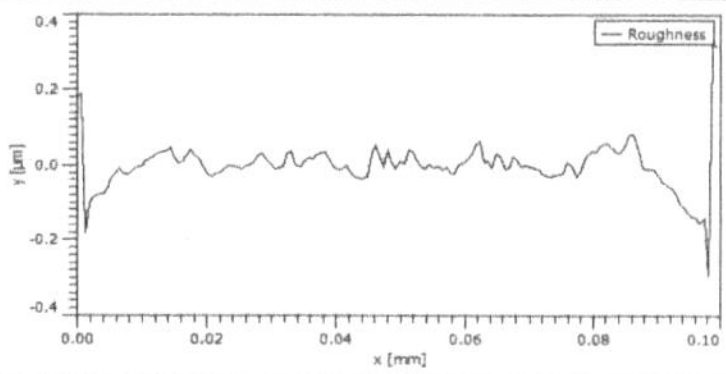
</td></tr>
<tr><td>Figure 4.12(c) [Exp. No.3]</td><td>Rendered image [Exp. No.3]</td><td>Ra 0.463 µm[Exp. No.2]</td></tr>
<tr><td>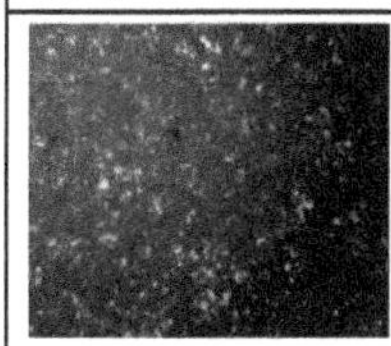</td><td>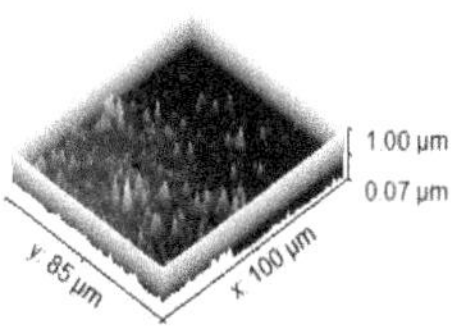
</td><td>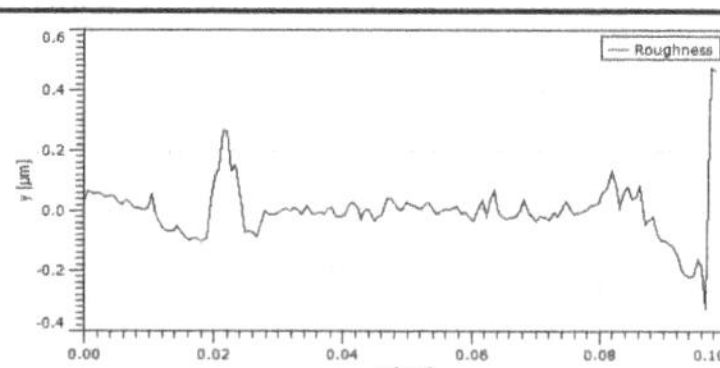
</td></tr>
<tr><td>Figure 4.12(d) [Exp. No.4]</td><td>Rendered image [Exp. No.4]</td><td>Ra 0.803 µm [Exp. No.4]</td></tr>
<tr><td></td><td>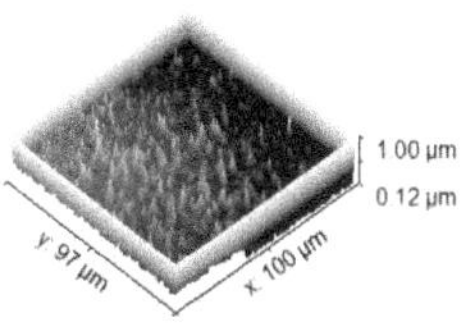
</td><td>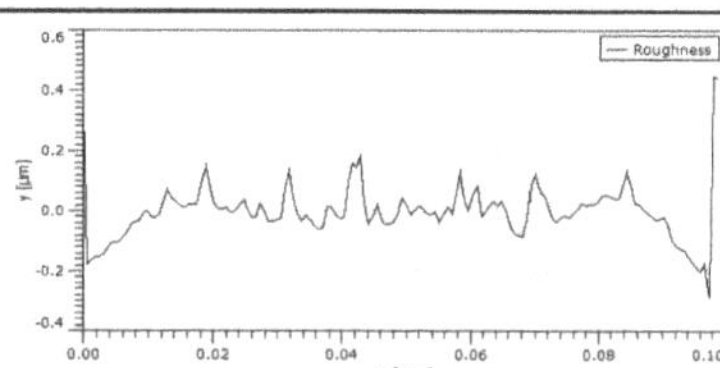
</td></tr>
<tr><td>Figure 4.12(e) [Exp. No.5]</td><td>Rendered image [Exp. No.5]</td><td>Ra 0.997 µm [Exp. No.5]</td></tr>
</table>

FIGURE 4.12 (a) [Exp. No.1] Rendered image [Exp. No.1] Ra 0.590 µm [Exp. No.1]. (b) [Exp. No.2] Rendered image [Exp. No.2] Ra 0.353 µm [Exp. No.2]. (c) [Exp. No.3] Rendered image [Exp. No.3] Ra 0.463 µm [Exp. No.2]. (d) [Exp. No.4] Rendered image [Exp. No.4] Ra 0.803 µm [Exp. No.4]. (e) [Exp. No.5] Rendered image [Exp. No.5] Ra 0.997 µm [Exp. No.5].

(Continued)

	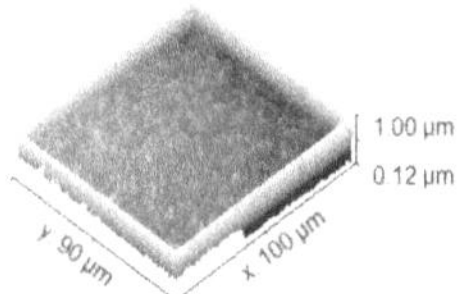	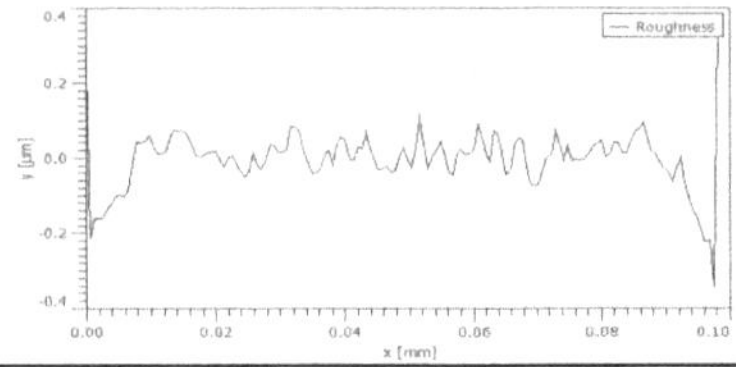
Figure 4.12(f) [Exp. No.6]	Rendered image [Exp. No.6]	Ra 0.630 µm [Exp. No.7]
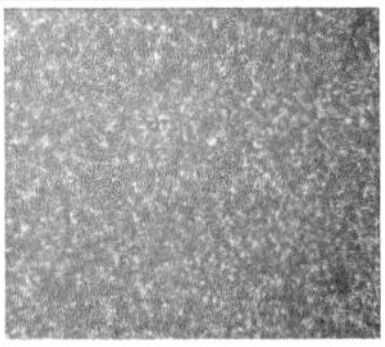	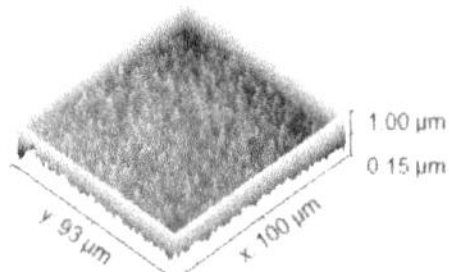	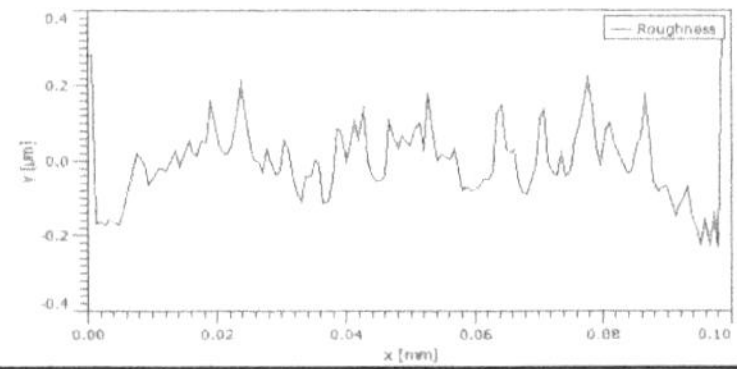
Figure 4.12(g) [Exp. No.7]	Rendered image [Exp. No.7]	Ra 0.537 µm [Exp. No.7]
	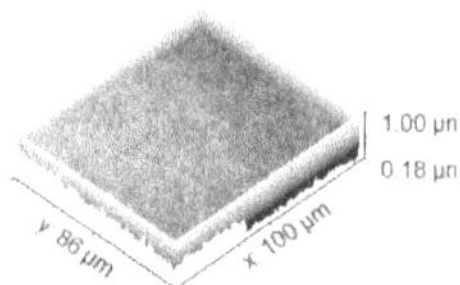	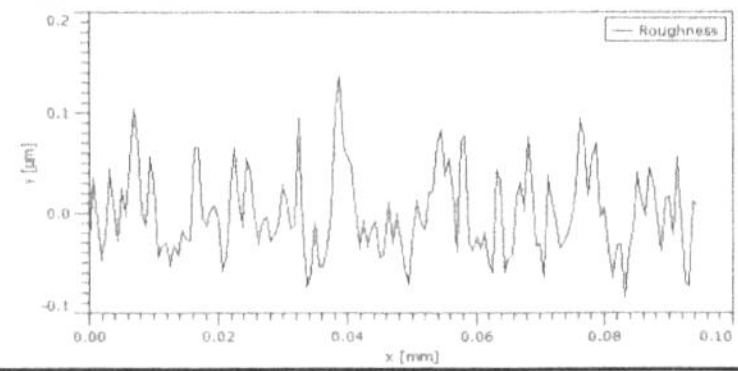
Figure4.12(h) [Exp. No.8]	Rendered image [Exp. No.8]	Ra 0.380 µm [Exp. No.8]
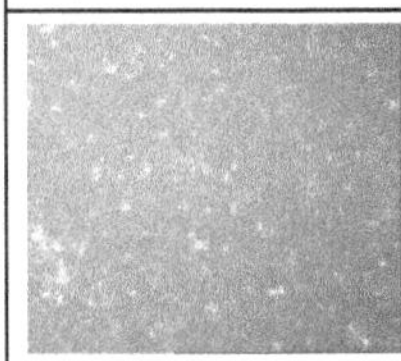	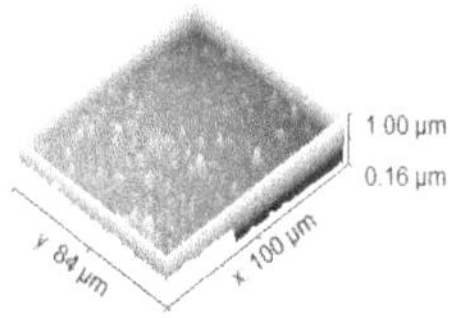	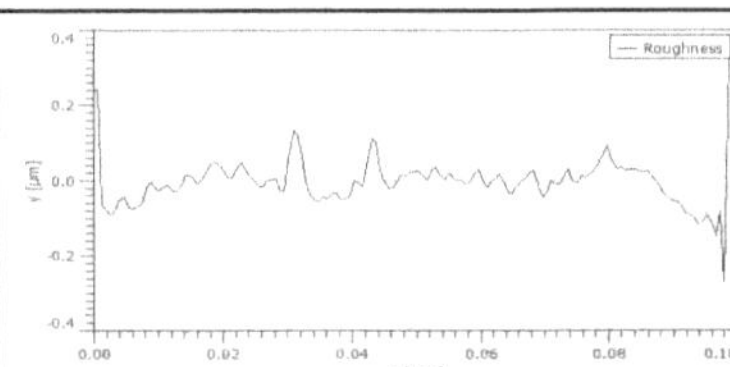
Figure 4.12(i) [Exp. No.9]	Rendered image [Exp. No.9]	Ra 0.673 µm [Exp. No.9]
	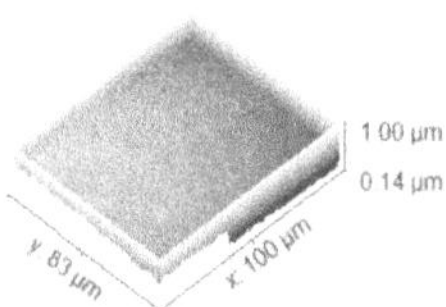	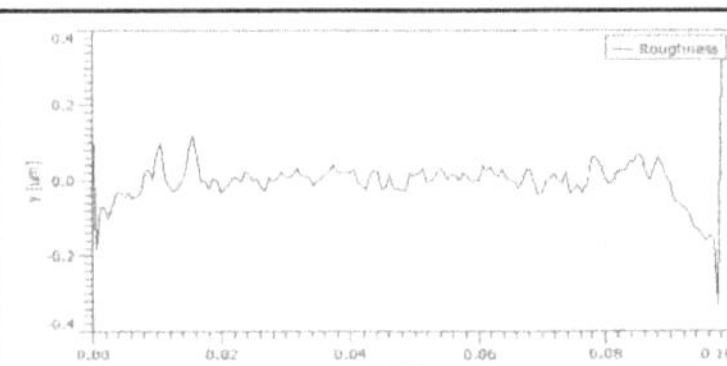
Figure 4.12(j) [Exp. No.10]	Rendered image [Exp. No.10]	Ra 0.863 µm [Exp. No.10]

FIGURE 4.12 (CONTINUED) (f) [Exp. No.6] Rendered image [Exp. No.6] Ra 0.630 µm [Exp. No.7]. (g) [Exp. No.7] Rendered image [Exp. No.7] Ra 0.537 µm [Exp. No.7]. (h) [Exp. No.8] Rendered image [Exp. No.8] Ra 0.380 µm [Exp. No.8]. (i) [Exp. No.9] Rendered image [Exp. No.9] Ra 0.673 µm [Exp. No.9]. (j) [Exp. No.10] Rendered image [Exp. No.10] Ra 0.863 µm [Exp. No.10].

(Continued)

	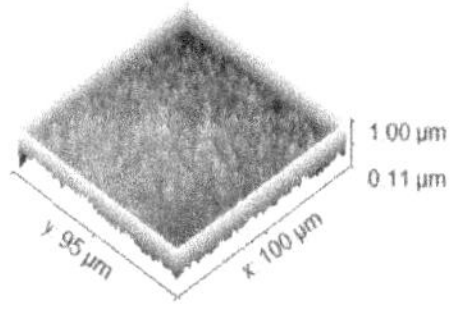	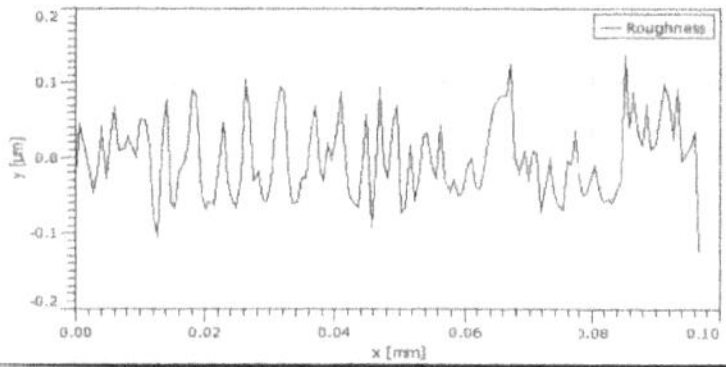
Figure4.12(k) [Exp.No.11]	Rendered image [Exp. No.11]	Ra 0.540 µm [Exp. No.11]
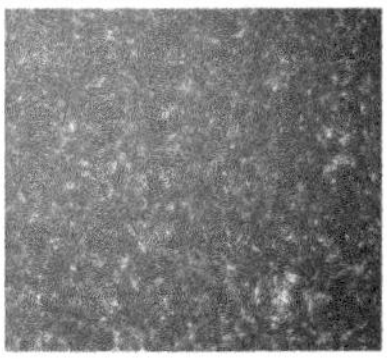	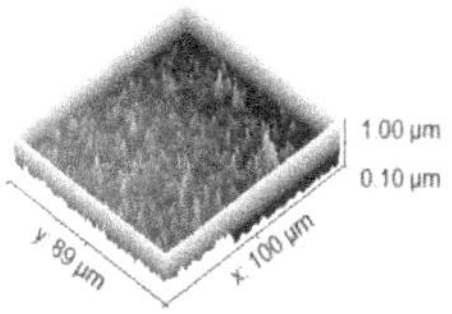	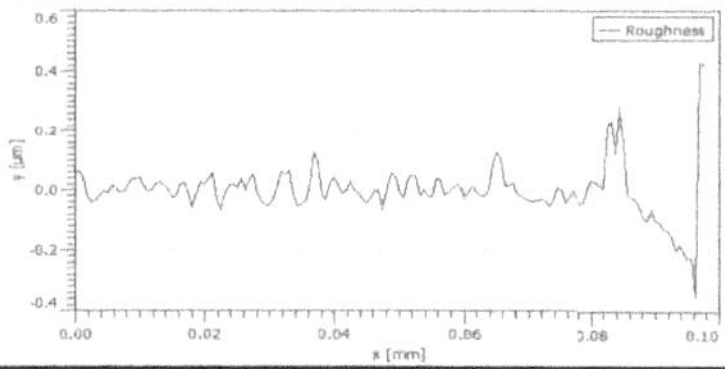
Figure 4.12(l) [Exp. No.12]	Rendered image [Exp. No.12]	Ra 1.310 µm [Exp. No.12]
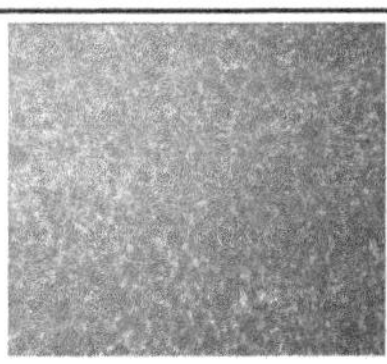	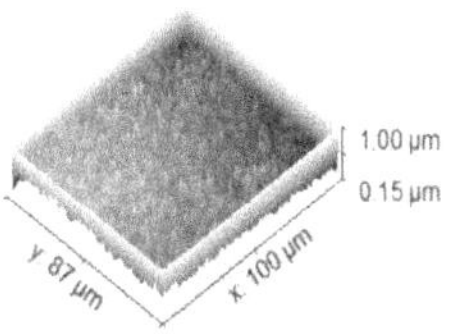	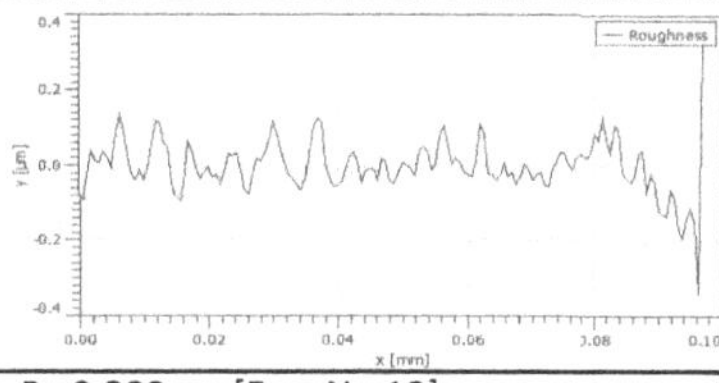
Figure 4.12(m) [Exp. No.13]	Rendered image [Exp. No.13]	Ra 0.800 µm[Exp. No.13]
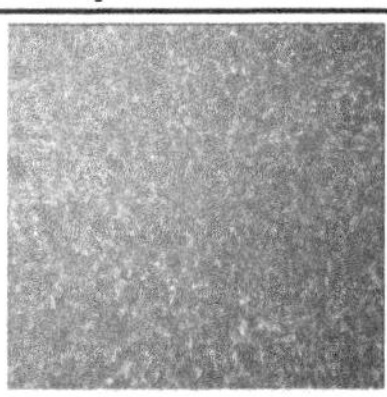	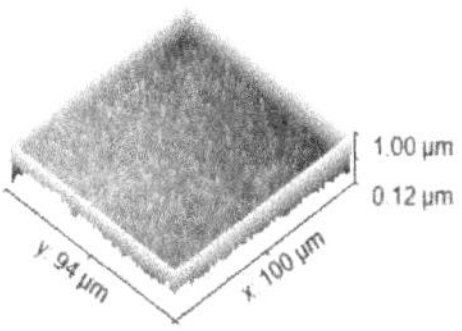	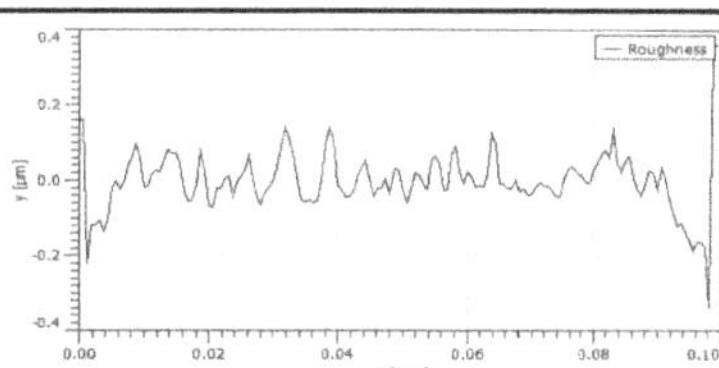
Figure 4.12(n) [Exp.No.14]	Rendered image [Exp. No.14]	Ra 0.730 µm[Exp. No.14]
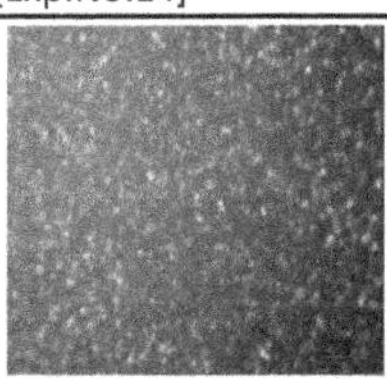	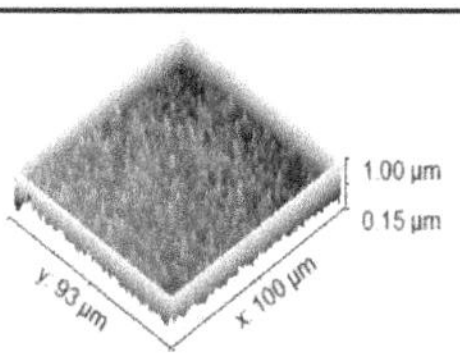	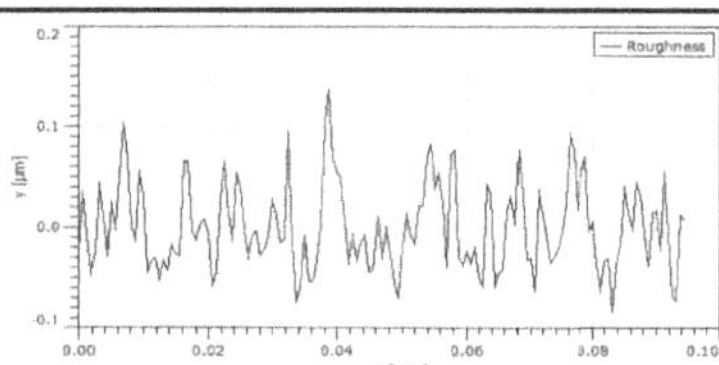
Figure 4.12(o) [Exp. No.15]	Rendered image [Exp. No.15]	Ra 0.880 µm [Exp. No.15]

FIGURE 4.12 (CONTINUED) (k) [Exp.No.11] Rendered image [Exp. No.11] Ra 0.540 µm [Exp. No.11]. (l) [Exp. No.12] Rendered image [Exp. No.12] Ra 1.310 µm [Exp. No.12]. (m) [Exp. No.13] Rendered image [Exp. No.13] Ra 0.800 µm [Exp. No.13]. (n) [Exp.No.14] Rendered image [Exp. No.14] Ra 0.730 µm [Exp. No.14]. (o) [Exp. No.15] Rendered image [Exp. No.15] Ra 0.880 µm [Exp. No.15].

(Continued)

	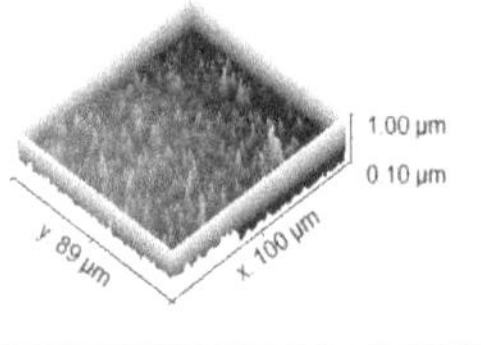	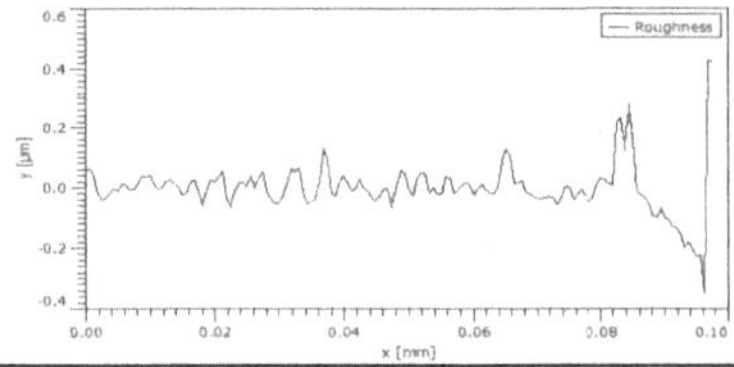
Figure 4.12(p) [Exp. No.16]	Rendered image [Exp. No.16]	Ra 0.947 µm[Exp. No.16]
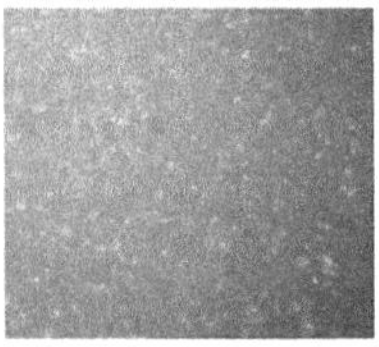	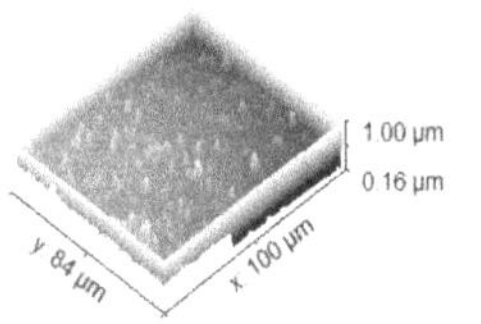	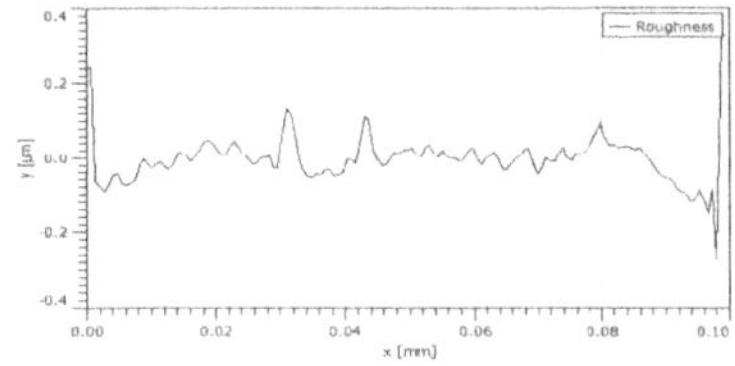
Figure 4.12(q) [Exp. No.17]	Rendered image [Exp. No.17]	Ra 0.990 µm[Exp. No.17]
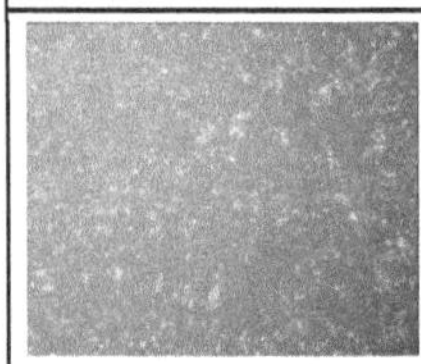	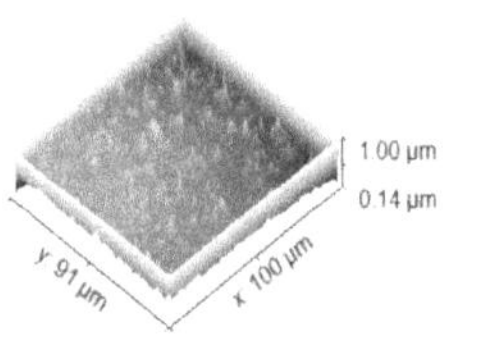	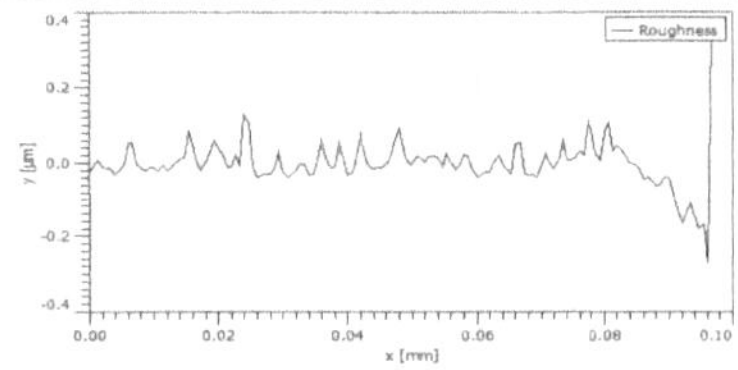
Figure 4.12(r) [Exp. No.18]	Rendered image [Exp. No.18]	Ra 0.473 µm [Exp. No.18]

FIGURE 4.12 (CONTINUED) (p) [Exp. No.16] Rendered image [Exp. No.16] Ra 0.947 µm [Exp. No.16]. (q) [Exp. No.17] Rendered image [Exp. No.17] Ra 0.990 µm [Exp. No.17]. (r) [Exp. No.18] Rendered image [Exp. No.18] Ra 0.473 µm [Exp. No.18].

finishing zone (5), followed by semifinishing zone (4), and superfinishing zone (3), and the fourth section is recess for the tool holder (1). The roughing zone comprises coarse grain with relatively high proportion of composite/abrasive particles in the polymer matrix. The semifinishing zone has relatively fine grains of abrasives with less proportion as compared to roughing zone. Finally, superfinishing zone has very fine abrasive particles with more proportion of the polymer matrix. Specifically, rough finishing zone (5) comprises acrylonitrile butadiene styrene (ABS)thermoplastic matrix reinforced with Al_2O_3/SiC of 200 µm size, semifinishing zone (4) basically comprises ABS thermoplastic matrix reinforced with Al_2O_3/SiC of 100 µm size, and superfinishing zone (3) comprises ABS thermoplastic matrix reinforced with Al_2O_3/SiC of 200 µm size. From the tool tip side, the tool enters into the workpiece through conventional ultrasonic drilling mechanism. As the tool progresses in the workpiece to generate cavity, the semifinishing and superfinishing zones of tool come in touch with the walls of hole, which actually improves the surface finish as a single stroke process.

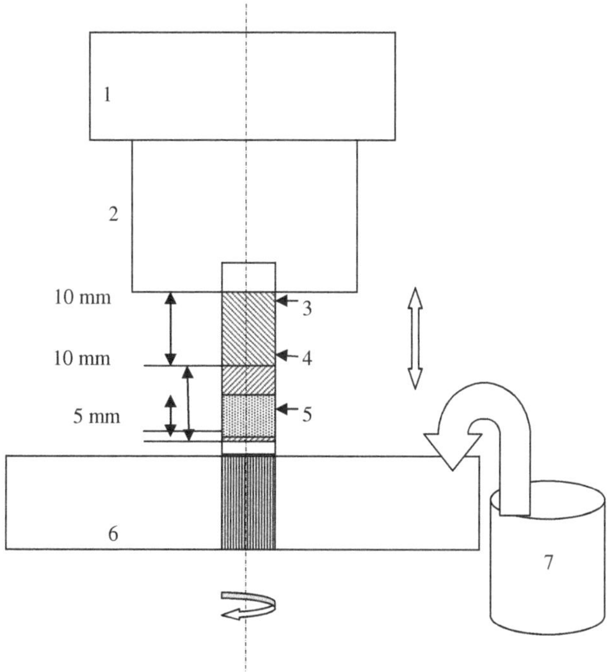

1	Transducer
2	Horn
3	ABS thermoplastic matrix reinforced with Al_2O_3/SiC of 50μm size
4	ABS thermoplastic matrix reinforced with Al_2O_3/SiC of 100μm size
5	ABS thermoplastic matrix reinforced with Al_2O_3/SiC of 200μm size
3+4+5	Rapid tooling
6	Workpiece (such as glass/Cast iron/ die steel etc.)
7	Slurry tank for supply of abrasive particles

FIGURE 4.13 USM-based broaching process.

4.6.2 Modified Ultrasonic Drilling (Without Use of Slurry) [63]

It should be noted that in the previous section, there was no direct contact of tool with workpiece, but in this case, the contact existed between the tool and the workpiece. However, the tool is made up of thermoplastic matrix, and the workpiece is in rotating condition. When the rapid tool comprising the polymer blended with abrasives (prepared through fused deposition modelling FDM) is allowed to just touch the workpiece, because of heat generation, the polymer matrix starts flowing, resulting in a strike of abrasive particles on the workpiece. This leads to the generation of cavity as in case of the conventional USM. The advantage of this technique is

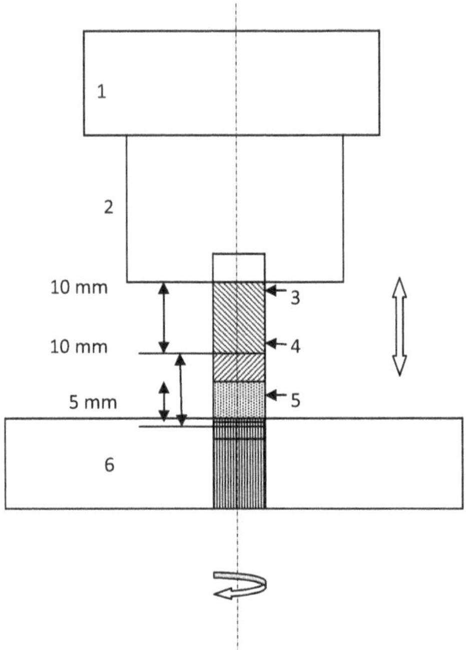

1	Transducer
2	Horn
3	ABS thermoplastic matrix reinforced with Al_2O_3/SiC of 50μm size
4	ABS thermoplastic matrix reinforced with Al_2O_3/SiC of 100μm size
5	ABS thermoplastic matrix reinforced with Al_2O_3/SiC of 200μm size
3+4+5	Rapid tooling
6	Workpiece (such as glass/Cast iron/ die steel etc.)

FIGURE 4.14 Modified ultrasonic drilling (without the use of slurry).

that it leads to great saving in slurry (i.e., no circulating water is required). The S.R of 150–200 nm at the cutoff length of 0.04 mm is possible for inner hole produced using this process. But tool life has been compromised significantly. But the overhead machining is possible. Figure 4.14 shows the possible hybridized process of USM without slurry application.

4.6.3 Ultrasonic Machining with Minimum Quantity Lubrication [63]

In this possible hybridized process, the water is provided (without slurry content) through drip at the tool–workpiece interface. The water droplets will act as a carrier for abrasive particles spilled out from the tool (because of heating of polymer matrix), thus resulting in better tool life. Figure 4.15 shows the possible working principle of USM with a minimum quantity of lubrication.

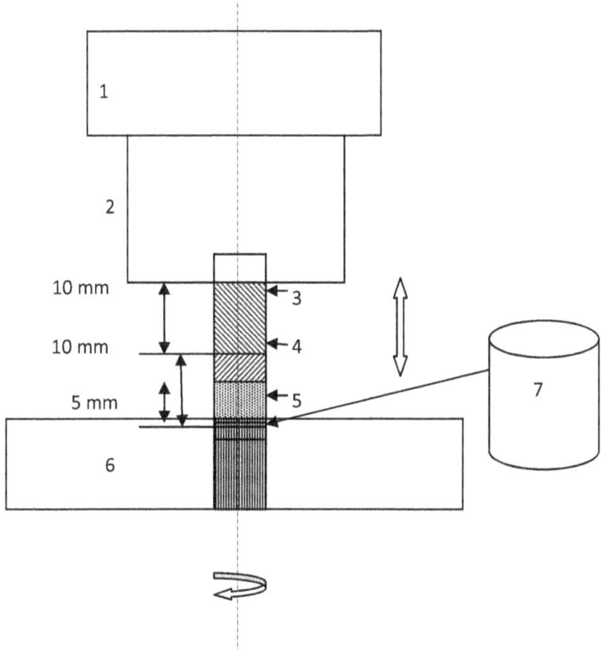

1	Transducer
2	Horn
3	ABS thermoplastic matrix reinforced with Al_2O_3/SiC of 50μm size
4	ABS thermoplastic matrix reinforced with Al_2O_3/SiC of 100μm size
5	ABS thermoplastic matrix reinforced with Al_2O_3/SiC of 200μm size
3+4+5	Rapid tooling
6	Workpiece (such as glass/Cast iron/ die steel etc.)
7	Water supply at tool tip with drip/ syringe for minimum quantity lubrication

FIGURE 4.15 USM with the minimum quantity lubrication.

4.6.4 Modified USM with Electric Discharge Machining with Minimum-Quantity Lubrication [63]

The process further provides a potential difference between the tool and the workpiece. The tool is made up of negatively charged surface, and the workpiece as positively charged surface. When the potential difference is given, the electrons start moving from tool (like in case of electric discharge machining) and strike the workpiece, which is positively charged. This results in an increase in the MRR. Figure 4.16 shows the suggested hybridized process of USM with electric discharge machining.

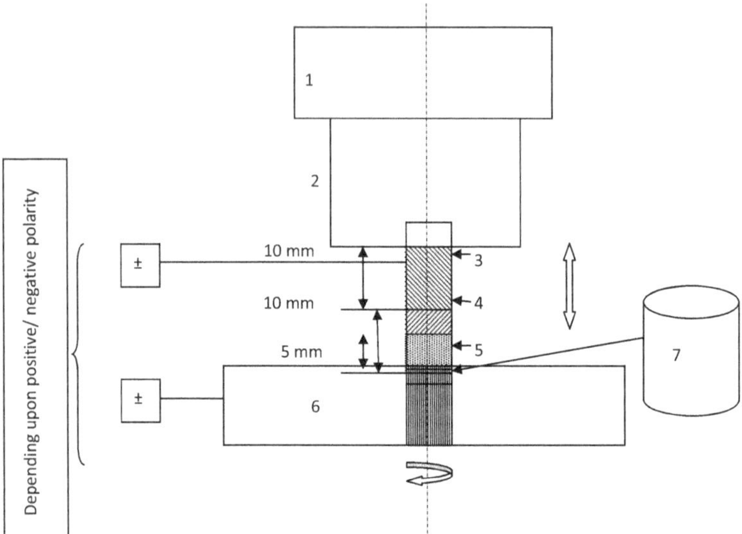

1	Transducer
2	Horn
3	ABS thermoplastic matrix reinforced with graphene, Al_2O_3/SiC of 50μm size
4	ABS thermoplastic matrix reinforced with graphene, Al_2O_3/SiC of 100μm size
5	ABS thermoplastic matrix reinforced with graphene, Al_2O_3/SiC of 200μm size
3+4+5	Rapid tooling set as negative terminal
6	Workpiece (die steel should be conducting) set as positive terminal
7	Water supply at tool tip with drip/ syringe for minimum quantity lubrication

FIGURE 4.16 Modified USM with electric discharge machining with the minimum-quantity lubrication.

4.6.5 Hybridization of Modified USM with Magnetic Abrasive Machining [63]

The USM tool may be made up of ferromagnetic reinforcements, and these tools can be used for the machining of ferromagnetic workpiece. Here under the influence of magnetic field (provided through electromagnet), the abrasive particles spilled out from the tool are getting attracted to the workpiece and strike there with high momentum, thus resulting in the formation of cavity as a mirror image of the tool. Figure 4.17 shows the suggested modified USM with magnetic abrasive machining.

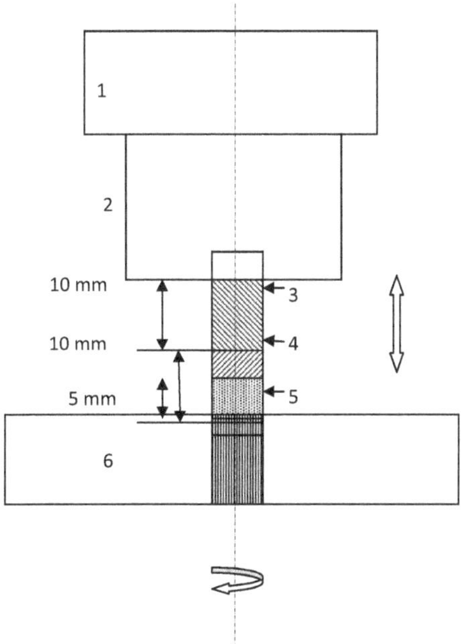

1	Transducer
2	Horn
3	ABS thermoplastic matrix reinforced with ferrous particles, Al_2O_3/SiC of 50µm size
4	ABS thermoplastic matrix reinforced with ferrous particles, Al_2O_3/SiC of 100µm size
5	ABS thermoplastic matrix reinforced with ferrous particles, Al_2O_3/SiC of 200µm size 3+4+5 Rapid tooling
6	Workpiece (magnetic in nature)

FIGURE 4.17 Hybridization of modified USM with magnetic abrasive machining.

4.7 CONCLUSIONS

This investigation of machining commercially Ti15 with different input parameters on USM setup has provided various observations, which may be helpful in machining the tough materials wisely by suitably choosing input parameters of USM machine. Further, in this research work, five possible cases of hybridization of USM have also been discussed. The observed critical points while machining the Ti15 are as follows:

1. From response surface optimization, it has been observed that for the three outputs together, the best settings were found to be T.M of level 6 (material-Ti), S.C of level 3 (25%), S.T of level 3 (alumina), S.Temp of level 1 (10°C),

P.R of level 1 (30), and S.G.S of level 3 (500 μm), thus playing a vital role in multioptimization of properties together (see Figure 4.8).

2. From contour plots of TWR taking P.R and S.Temp as input parameters, it has been observed that the maximum wear rate has been obtained for 20 < S.Temp < 50 range and for P.R that was found to be less than 33 for the nominal results (see Figure 4.10). When MRR as output and S.Temp (*x*-axis) and P.R (*y*-axis) as input conditions were analyzed by contour plot, it was observed that S.Temp in the range of 22°C–55°C and P.R greater than 70 have provided the maximum MRR. When S.R was taken as output property for the same input parameters for contour plotting, it has been observed that the range of 15 < S.Temp < 35 and 40 < P.R < 70 showed the maximum contribution towards S.R.

In this investigation, five different possible hybridized USM machining processes have been discussed, which may have advantages over the traditional processes of machining; moreover, the hybridized process of USM processing may provide better solutions to industries in terms of processing and cost.

ACKNOWLEDGMENT

The authors are highly obliged to Manufacturing Research Laboratory, Guru Nanak Dev Engineering College (GNDEC), Ludhiana, and Thapar Institute of Engineering and Technology, Patiala, for providing laboratory facilities and other technical supports.

REFERENCES

1. F. Benedict Gary, *Book on Non Traditional Manufacturing Processes*, Marcel Dekker, Inc, New York, 1987, pp. 67–86 [Chapter 6].
2. R. Gilmore, Ultrasonic machining of ceramics, SME Paper MS90-346, 1990, p. 12.
3. L.D. Rozenberg, V.F. Kazantsev, *Ultrasonic Cutting*, Consultants Bureau, New York, 1964.
4. J.B. Kohls, Ultrasonic-manufacturing process: ultrasonic machining (USM) and ultrasonic impact grinding (US1G), *Carbide Tool J.* 16 (5) (1984) 12–15.
5. M. Haslehurst, *Manufacturing Technology*, 3rd ed., Hodder and Stoughton, London, 1981, pp. 270–271.
6. V. Soundararajan, V. Radhakrishnan, An experimental investigation on the basic mechanisms involved in ultrasonic machining, *Int. J. MTDR* 26 (3) (1986) 307–321.
7. A. Satyanarayana, B.G. Krishna Reddy, Design of velocity transformers for ultrasonic machining, *Electrical India* 24 (14) (1984) 11–20.
8. T.J. Drozda, C. Wick, Non-traditional machining, Book Chapter 29: *Tool and Manufacturing Engineers Handbook* (Desk Ed.), vol. 1, Society of Manufacturing Engineers, Dearborn, MI, 1983, pp. 1–23.
9. G. Nishimura, Ultrasonic machining, part I, *J. Fac. Eng. Tokyo Univ.* 24 (3) (1954) 65–100.
10. E.A. Neppiras, Report on ultrasonic machining, *Metalwork. Prod.* 100 (1956) 1283–1288, 1333–1336, 1377–1382, 1420–1424, 1464–1468, 1554–1560, 1599–1604.
11. E.J. Weller, *Non-traditional Machining Processes*, 2nd ed., Society of Manufacturing Engineers, Dearborn, MI, 1984, pp. 15–71.

12. J.O. Fairer, English Patent No. 602801 from 3 June 1948—USM.
13. T.B. Thoe, D.K. Aspinwall, M.L.H. Wise, Review on ultrasonic machining, *Int. J. Mach. Tools Manuf.* 38 (4) (1998) 239–255.
14. K.H.W. Scab, Parametric studies of ultrasonic machining, SME Tech. Paper MR90-294, 1990, p. 11.
15. E.A. Neppiras, Macroson. *Ind.: Ultrasonics* 10 (1972) 9–13.
16. J. Perkins, An outline of power ultrasonic, Technical Report by Kerry Ultrasonics, 1972, p. 7.
17. F.T. Farago, *Abrasive Methods Engineering*, vol. 2, Industrial Press, New York, 1980, pp. 480–481.
18. L. Balamuth, Ultrasonic vibrations assist cutting tools, *Metalwork. Prod.* 108 (24) (1964) 75–77.
19. D.C. Kennedy, R.J. Grieve, Ultrasonic machining—a review, *Prod. Eng.* 54 (9) (1975) 481–486.
20. D. Kremer, New developments on ultrasonic machining, SME Technical Paper MR91-522, 1991, p. 13.
21. D. Clifton, Y. Imal, J.A. Mc-Geough, Some ultrasonic effects on machining materials encountered in the offshore industries, in *Proceedings of the 30th International MATADOR Conference*, 1993, pp. 119–123.
22. M.C. Shaw, Ultrasonic grinding, *Microtechnic* 10(6) (1956) 257–265.
23. M.A. Moreland, Ultrasonic machining—Book Chapter: Ceramics and glasses. in: J. Schneider, Samuel (Eds.), *Engineering Material Handbook*, vol. 4, ASM International, Cleveland, OH, 1991, pp. 359–362.
24. P.L. Guzzo, A.H. Shinohara, A.A. Raslan, A comparative study on ultrasonic machining of hard and brittle materials, *J. Brazil Soc. Mech. Sci. Eng.* 26 (1) (2004) 56–61.
25. M. Komaraiah, M.A. Manan, P. Narasimha Reddy, S. Victor, Investigation of surface roughness and accuracy in ultrasonic machining, *Precis. Eng.* 10 (2) (1988) 59–65.
26. M. Komaraiah, P. Narasimha Reddy, A study on the influence of workpiece properties in ultrasonic machining, *Int. J. Mach. Tools Manuf.* 33 (3) (1993) 495–505.
27. Z.C. Li, Y. Jiao, T.W. Deines, Z.J. Pei, C. Treadwell, Rotary ultrasonic machining of ceramic matrix composites: feasibility study and designed experiments, *Int. J. Mach. Tools Manuf.* 45 (12–13) (2005) 1402–1411.
28. K.H.W. Seah, Y.S. Wong, L.C. Lee, Design of tool holders for ultrasonic machining using FEM, *J. Mater. Process. Technol.* 37 (1–4) (1993) 801–816.
29. E.A. Neppiras, Ultrasonic machining and forming, *Ultrasonics* 2 (1964) 167–173.
30. A.C. Wang, B.H. Yan, X.T. Li, F.Y. Huang, Use of micro ultrasonic vibration lapping to enhance the precision of microholes drilled by micro electro-discharge machining, *Int. J. Mach. Tools Manuf.* 42 (8) (2002) 915–923.
31. Z.C. Li, Y. Jiao, T.W. Deines, Z.J. Pei, C. Treadwell, Rotary ultrasonic machining of ceramic matrix composites: feasibility study and designed experiments, *Int. J. Mach. Tools Manuf.* 45 (12–13) (2005) 1402–1411.
32. L.A. Balamuth, Ultrasonic assistance to conventional metal removal, *Ultrasonics* 4 (1966) 125–130.
33. A.I. Isaev, Learning with ultrasonically vibrated reamers, *Mach. Tooling* 33 (6) (1962) 27–30.
34. R. Singh, J.S. Khamba, Silver brazing for tool preparation in USM process, in *Proceedings of the National Workshop of Welding Technology in India—Present Status and Future Trends*, SLIET Longowal (Pb.) India, 2003, pp. 61–63.
35. R. Singh, J.S. Khamba, Tool manufacturing technique in ultrasonic drilling machine, *J. Manuf. Technol. Today* 3 (1) (2004) 5–7.

36. Compiled by the Technical Staff of the Machinability Data Centre, *Machining Data Handbook*, 3rd ed., vol. 2, Cincinnati Metcut Research Associates Inc., Cincinnati, OH, 1980, pp. 43–63.
37. R. Halm, P. Schulz, Ultrasonic machining of complex ceramic components, Erosion AC Report, DKG 70, No. 7, 1993, p. 6.
38. K.F. Graff, Macrosonics in industry. 5. Ultrasonic machining, *Ultrasonics* 13 (1975) 103–109.
39. M.A. Moreland, Ultrasonic advantages revealed in the hole story, *Ceram. Appl. Manuf.* 187 (1988) 156–162.
40. R. Gilmore, Ultrasonic machining and orbital abrasion techniques, SME Technical Paper (Series) AIR, NM89-419, 1989, pp. 1–20.
41. D. Moore, Ultrasonic impact grinding, in *Proceedings of the Nontraditional Machining Conference, Cincinnati*, 1985, pp. 137–139.
42. P. Black, An ultrasonic impact grinding technique for electrode forming and redressing, in *Proceedings of the Non-traditional Machining Conference*, Cincinnati, Ohio, ASM, 1985, pp. 129–136.
43. D. Kremer, G. Bazine, A. Moison, Ultrasonic machining improves EDM technology, in: J.R. Crookall (Ed.), *Proceedings of the Seventh International Symposium on Electro Machining*, Birmingham, UK, Kempston, Bedford : IFS (Publications), 1983, pp. 67–76.
44. S.R. Ghabrial, Trends towards improving surfaces produced by modem processes, in *Paper presented at the Third International Conference on Metrol and Prop, of Eng'g Surf*, Teesside, England, 1986, pp. 113–118.
45. M.W. Robare, D.W. Richerson, Proceedings of the ARPA/NAVSEAGarrett/Ai Research Ceramic Gas Turbine Engine Demonstration Program Review at Rotor Blade Machining Development, Marine Maritime Academy, 1977.
46. R. Singh, Ultrasonic machining for tough materials and its application in mechanical industry, in *Proceedings of the Fourth National Symposium of Research Scholars on Metal and Materials*, IIT Madras (India), 2002, p. 31.
47. D.R.S.V. Verma, B.G. Nanda Gopal, K. Srinivasulu, S. Sudhakar Reddy, Effect of pre-drilled holes on tool life in turning of aerospace titanium alloys, AMS-03, in *Proceedings of the National Conference on Advances in Manufacturing System*, Production Engineering Department, Jadavpur University, Kolkata, India, 2003, pp. 42–47.
48. D.A. Dornfeld, J.S. Kim, H. Dechow, J. Hewsow, L.J. Chen, Drilling burr formation in titanium alloy Ti–6Al–4V, *Ann. CIRP* 48 (1) (1999) 73–76.
49. C. Wick. Tool and Manufacturing Engineers Handbook: Materials, Finishing and Coating (Vol. 3). Society of manufacturing Engineers, 1985.
50. J. Khamba, R. Singh, Effect of alumina (white fused) slurry in ultrasonic assisted drilling of titanium alloys (TITAN 15), in *Proceedings of the National Conference on Materials and Related Technologies (NCMRT-2003)* at TIET Patiala (Pb.), India, 2003, pp. 75–79.
51. R. Singh, J.S. Khamba, A frame work for modeling the machining characteristics of titanium alloys using USM, in *Proceedings of the International Conference on Digital-aided Modeling and Simulation at CIT*, Coimbatore, India, 2003, p. 31.
52. A.L. Mantle, D.K. Aspinwall, Single point turning of titanium aluminide intermetallic, in *Titanium 95, Proceedings of the Eighth World Conference on Titanium*, vol. 1, 1995, pp. 248–255.
53. R. Singh, J.S. Khamba, 2009. Mathematical modeling of tool wear rate in ultrasonic machining of titanium. *Int. J. Adv. Manufact. Technol.* 43(5–6), pp. 573–580.
54. R. Singh, 2010. Comparison of statistically controlled machining solutions of titanium alloys using USM. *Int. J. Automot. Mech. Eng.* 1(1), pp. 66–78.

55. R. Singh, J.S. Khamba, 2008. Comparison of slurry effect on machining characteristics of titanium in ultrasonic drilling. *J. Mater. Process. Tech.* 197(1–3), pp. 200–205.
56. R. Singh, 2010. Study of statistically controlled surface roughness solution in machining of titanium alloys using ultrasonic machining. *Int. J. Eng. Syst. Modell. Simul.* 2(3), pp. 149–153.
57. R. Singh, 2010. Comparison of statistically controlled machining solutions of titanium alloys using USM. *Int. J. Automot. Mech. Eng.* 1(1), pp. 66–78.
58. G.K. Dhuria, R. Singh, A. Batish, 2011. Ultrasonic machining of titanium and its alloys: a state of art review and future prospective. *Int. J. Mach. Machin. Mater.* 10(4), pp. 326–355.
59. G. Dhuria, R. Singh, A. Batish, 2016. Predictive modeling of surface roughness in ultrasonic machining of cryogenic treated Ti-6Al-4V. *Eng. Comput.* 33(8), pp. 2377–2394.
60. G.K. Dhuria, R. Singh, A. Batish, 2017. Application of a hybrid Taguchi-entropy weight-based GRA method to optimize and neural network approach to predict the machining responses in ultrasonic machining of Ti–6Al–4V. *J. Brazil. Soc. Mech. Sci. Eng.* 39(7), pp. 2619–2634.
61. R. Singh, J.S. Khamba, 2007. Taguchi technique for modeling material removal rate in ultrasonic machining of titanium. *Mater. Sci. Eng.: A* 460, pp. 365–369.
62. R. Singh, J.S. Khamba, 2007. Macromodel for ultrasonic machining of titanium and its alloys: designed experiments. *Proc. Inst. Mech. Eng., Part B* 221(2), pp. 221–229.
63. Modified ultrasonic machining process (Ref. No. Indian Patent Application No: 202011001505 dated January 13, 2020, TEMP/E-1/1644/2020-DEL). Inventor: Dr. Rupinder Singh.

5 A Trending Nonconventional Hybrid Finishing/ Machining Process

Ankit Sharma
Chitkara University

Vishwas Grover
Ajay Kumar Garg Engineering College

Atul Babbar
Thapar Institute of Engineering and Technology, Patiala

Richa Rani
Panjab University

CONTENTS

5.1 INTRODUCTION

In today's world, with an increasing demand of highly precise products, there is a requirement of hybrid machining processes. Hybrid machining processes are the combination of machining processes to have more efficient and productive part [1,2].

In the hybrid machining processes, various energy forms are acting over the identical area at identical time for machining, which makes these processes a little bit complex. These processes have very wide applications in machining various machine tools. These processes can be used in electronics, automotive, and aerospace industries where high accuracy and surface quality are of great concern [3–5]. Different hybrid machining processes are being used for different purposes. Some hybrid machining processes are used for machining such as turning or drilling, but certain hybrid machining processes are used for finishing purposes. In today's industry, highly precise finished industrial products are of great demand. Various hybrid machining processes for surface finishing have been evolved in recent years for obtaining precisely finished industrial products with good surface characteristics over their internal surfaces.

5.2 HYBRID MACHINING PROCESSES FOR SURFACE FINISHING

Surface characteristics are highly demanding for increased operational life of different industrial products such as cylinder liner, air bearing, and die cavity. Such products require nano-finishing with close dimensional accuracy. Conventional finishing processes such as honing and grinding have been carried out to finish the surface of these components. These processes use a solid tool consisting of bounded abrasive particles. The solid tool provides uncontrolled finishing forces to the product's surface, which results in damage to the surface. These problems can be resolved using magnetorheological (MR) finishing processes. These processes use MR polishing fluids which behave like semisolid fluid under the magnetic field with controlled finishing forces. This context emphasizes on the capability of various magnetically assisted finishing processes for various internal and external surfaces with their mechanism of material removal.

A good surface quality is required in today's industries which mainly depend on the finishing method [5–11]. Various industrial products require mirror-like surface finish for their proper functionality and durability [12,13]. The surface quality can be enhanced for increasing the component's performance [14,15]. Therefore, the conventional finishing processes such as grinding and honing are mainly applied to finish the industrial components [16,17]. These processes use abrasive stones/sticks to finish the component's surfaces [18–22]. Abrasive stones/sticks approach the workpiece's surface with least control over the finishing forces, which leads to surface defects such as scratches, cavity, and grooves [23,24]. Past researchers [25] evaluated the heat flux produced when the rotating grinding wheel touches the workpiece surface. Another researcher [26] monitored the different grinding defects generated by the wrong wheel dressing. In a study, it was disclosed that the torn and folded metals are produced due to rigid abrasive sticks of the honing tool [23].

Further, the hybrid finishing processes such as electrochemical grinding (ECG) and electrochemical honing (ECH) have been developed for enhancing the performance of both grinding and honing processes [27,28]. The above conventional finishing processes are less effective because of their tool design structure.

However, it is challenging as the finishing forces applied by the tool are uncontrollable [29]. A number of nonconventional hybrid finishing processes have been developed by researchers to enhance the performance of the above-said methods. Various advanced hybrid machining processes with controlled finishing forces are stated below.

5.2.1 Magnetorheological Abrasive Flow Finishing (MRAFF) Process

MRAFF process has been developed for the internal cylindrical finishing of non-ferromagnetic materials, as represented in Figure 5.1a [30]. The material removal phenomenon in MRAFF process is represented in Figure 5.1bandc. In Figure 5.1b, the ferromagnetic material (e.g., mild steel, EN31) has been taken for finishing by MRAFF process. The higher magnetic gradient is found on the ferromagnetic surface than on the polishing fluid. Instead of finishing the internal cylindrical surface, the iron particles can remain attached to the workpiece surface only. It may result in the surface damages such as scratches and grooves [31]. In Figure 5.1c, the non-ferromagnetic material (e.g., copper and aluminum) has been taken for finishing with MRAFF process. The lesser magnetic field gradient is found on the non-ferromagnetic material than on the polishing fluid. The iron particles form a chain-like structure as per the magnetic flux gradient. These iron particles' chains help to roll the abrasive particles over the internal cylindrical surfaces [32]. The different forces act over the single abrasive particle, as represented in Figure 5.1c. The shear force (F_s) induced during MRAFF process can be calculated as [33]:

$$F_s = (A - A_1) \tag{5.1}$$

where A is the entire area of abrasive, and A_1 is the indented area abrasive.

The indented depth (d) of abrasive can be calculated as [33]:

$$d = \frac{d_a}{2} - \frac{1}{2}\sqrt{d_a^2 - d_i^2} \tag{5.2}$$

The indented diameter (d_i) of abrasive can be calculated as [33]:

$$d_i = \sqrt{d_a^2 + \left(d_a - \frac{2 \times 10^{-6} F_n}{9.81 H_{BHN} \Pi d_a}\right)^2} \tag{5.3}$$

where d_a is the abrasive particle diameter, F_n is the normal force utilized by the abrasive particle, and H_{BHN} is the Brinell hardness number.

The researcher considered the influence of magnetic field gradient on the surface roughness, and the results were found superior with the rise in magnetic field gradient [30]. In further studies, it was considered that there were the consequences of extrusion pressure and the quantity of finishing cycles on the final R_a value on the finished surface [34].

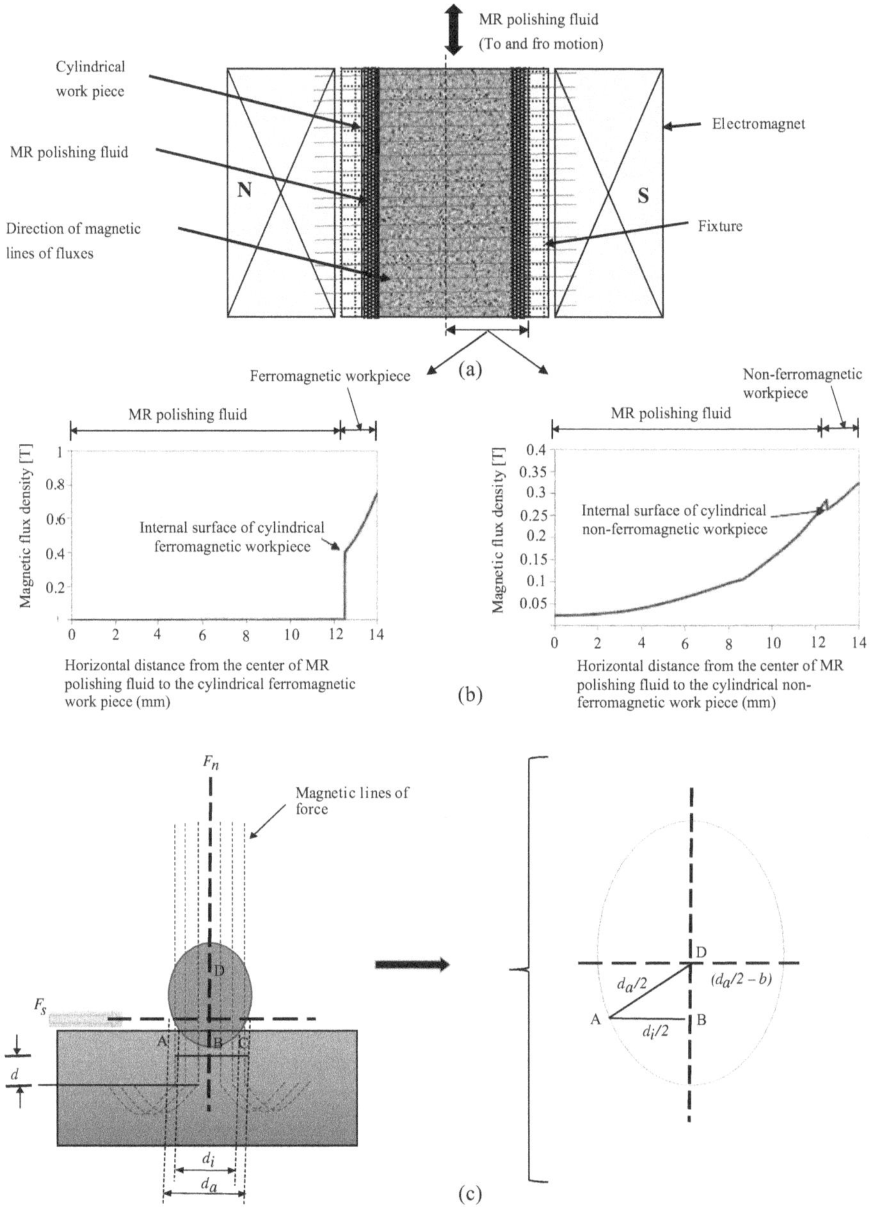

FIGURE 5.1 (a) Schematic of MRAFF process, magnetostatic field report of MRAFF process in finishing of the internal cylindrical specimen surface made up of (b) ferromagnetic specimen and (c) non-ferromagnetic specimen, and different forces acting on the single abrasive particle during MRAFF process [32, 33].

5.2.2 Rotational-Magnetorheological Abrasive Flow Finishing (R-MRAFF) Process

MRAFF process was slightly modified to increase the process performance and to raise the shearing tendency of the abrasive particles for enhancing the surface roughness. Therefore, it was named as R-MRAFF process, as shown in Figure 5.2a [35].

While finishing, the reciprocation movement of the piston delivers a vertical motion to the MR polishing fluid, whereas the revolving motion of the magnets provides the abrasive particles to roll on the internal periphery of the specimen. This raises the shearing capacity of abrasive particles for producing a precise surface finish. The magnetic field contours and vectors from magnetostatic simulations for R-MRAFF during finishing of non-ferromagnetic and ferromagnetic workpieces are represented in Figure 5.2b and c, respectively.

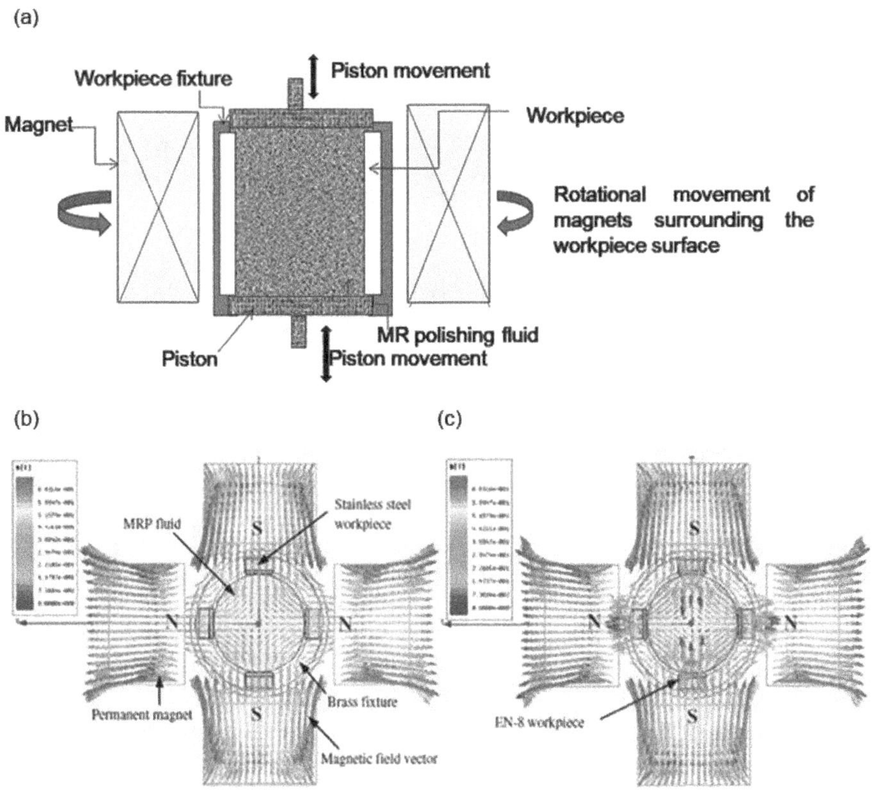

FIGURE 5.2 (a) Schematic of R-MRAFF, magnetostatic flux contour, and magnetic field vector of MRAFF process during finishing of (b) non-ferromagnetic workpiece and (c) ferromagnetic workpiece [35, 36].

As per simulation description, the R-MRAFF process is suitable for finishing of non-ferromagnetic workpieces because of the better allocation of magnetic field. In MRAFF process, the different forces induced are shown in Figure 5.3.

Similar to MRAFF process, the normal force (F_n) can be calculated as [36]:

$$F_n = k \times p \tag{5.4}$$

where k is the proportionality constant, and P is the extrusion pressure.

The magnetic force (F_m) induced by the iron particle due to the magnetic field gradient is calculated as [36]:

$$F_m = m \times \chi_m \times \mu \times H \times \nabla H \tag{5.5}$$

where m is the individual CIP mass, χ_m is the iron particle magnetic susceptibility, μ_0 is the free space magnetic permeability, H is the intensity of magnetic field, and ∇H is the gradient of the intensity of magnetic field.

Now, because of rotating magnets, the abrasive particles are rolled over the internal non-ferromagnetic material. Hence, the centrifugal force (F_{cen}) is induced and can be calculated as [36]:

$$F_{cen} = m \times r \times \omega^2 \tag{5.6}$$

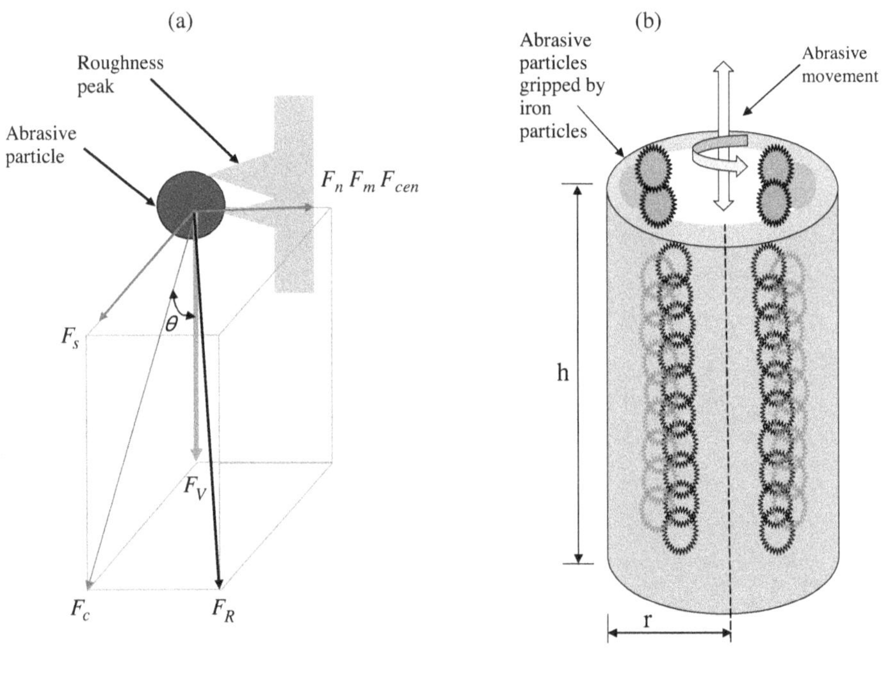

FIGURE 5.3 (a) Different forces acting on the abrasive particle at the time of finishing with R-MRAFF process and (b) abrasive particles moved on the internal cylindrical periphery of non-ferromagnetic specimen [36].

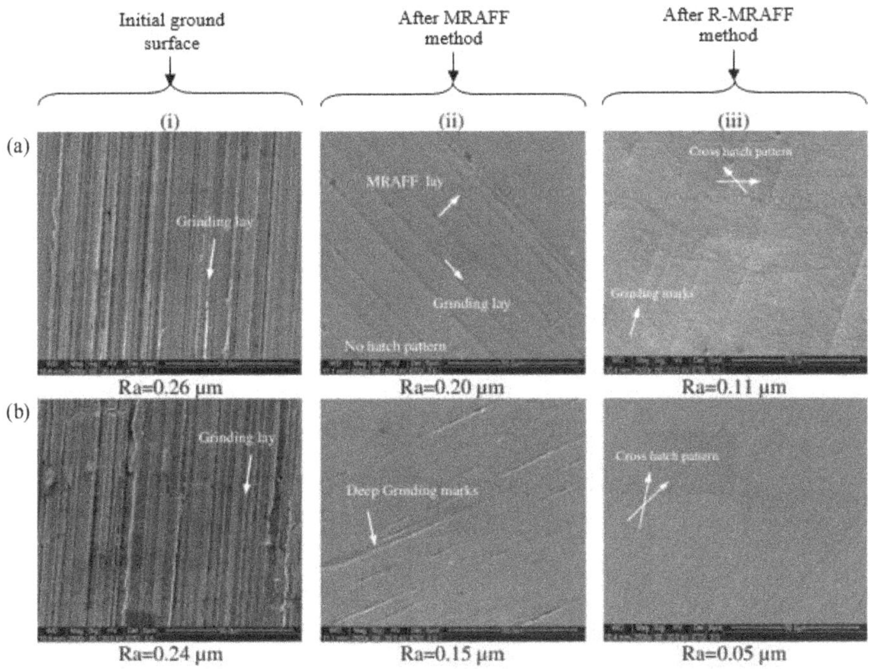

FIGURE 5.4 Surface morphology using scanning electron microscopy (SEM) images of (a) stainless steel workpiece and (b) brass workpiece [36].

The shearing force (F_s) induced by the abrasive particles due to the rotation of magnets can be calculated as [36]:

$$F_s = 2 \times m(\vec{v} \times \vec{\omega}) \tag{5.7}$$

Das et al. compared the finishing efficiency of MRAFF and R-MRAFF processes and found that the results of R-MRAFF process were found better, as shown in Figure 5.4 [36].

5.2.3 Ball-End Magnetorheological Finishing (BEMRF) Process

MRF and MR jet finishing processes are restricted to the particular geometries only because of the constraint on the relative motion of MR polishing fluid and the workpiece surfaces. Further, a novel finishing process is investigated for finishing the 3-D surfaces through ball-end MR finishing (BEMRF) tool, as shown in Figure 5.5 [37]. The process is named as ball-end MR finishing (BEMRF) process. In this process, MR polishing fluid enters from the top of the tool without applying the magnetic field. As soon as MR polishing fluid comes to the tip of the tool, the magnetic field is provided due to which MR polishing fluid gets stiffened. During finishing, the rotation and feed motion are provided by the tool to the workpiece surface.

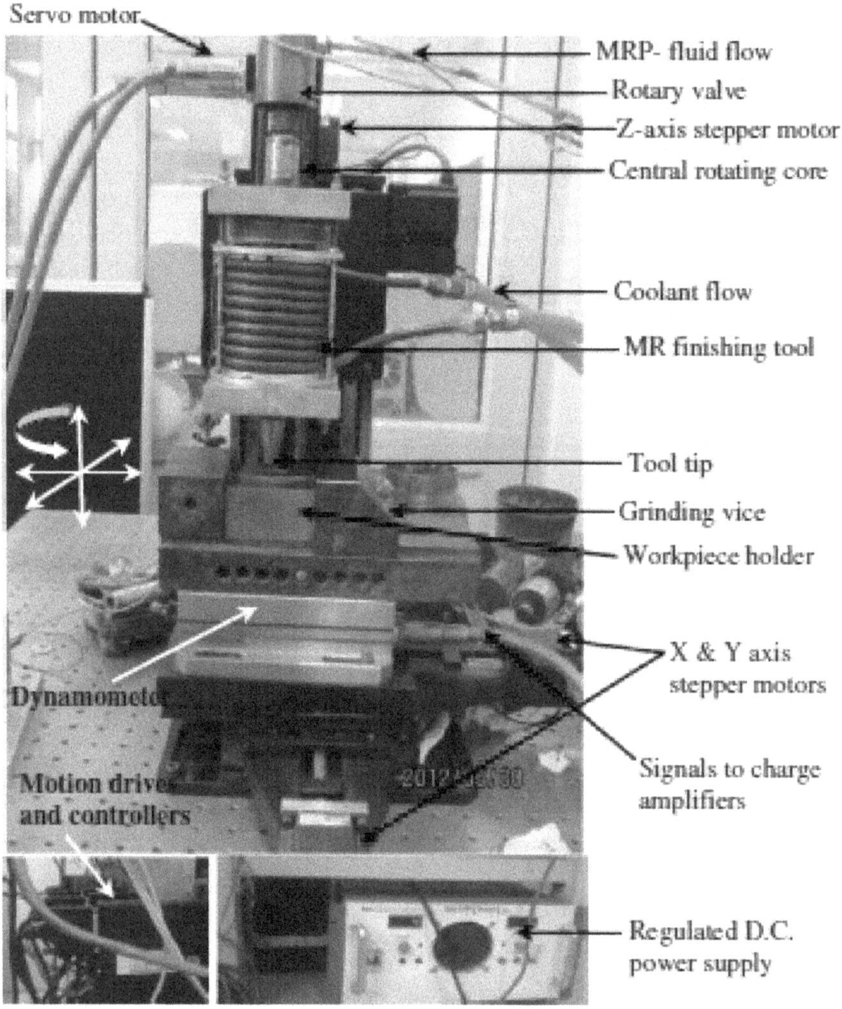

FIGURE 5.5 Photograph of BEMRF process [32].

The material removal mechanism in BEMRF process is represented in Figure 5.6. The abrasives get indent inside workpiece peaks owing to the lower magnetic field at the targeted periphery of the workpiece, as represented in Figure 5.6.

Due to the rotation of the central rotating core and relative motion between the abrasives and the specimen surface, indented abrasives break the roughness peaks into microchips, as shown in Figure 5.6b. Due to a constant feed of workpiece, all the peaks get removed from the surface and better-finished surface achieved, as shown in Figure 5.6c. By the action of shearing, gripped abrasives pluck out the workpiece material from its surface roughness. It is considered to finish flat as well as 3-D curved surfaces where fine finishing is greatly concerned [38].

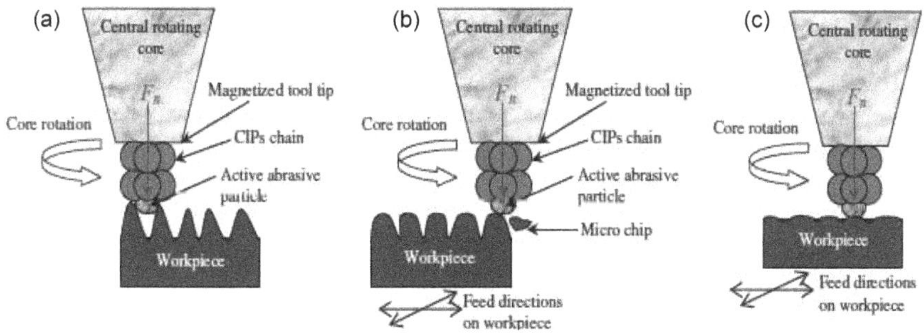

FIGURE 5.6 Material removal mechanism in the BEMRF process [38].

5.2.4 Magnetorheological Honing (MRH) Process

The MRH process has been developed for finishing of internal surface of ferromagnetic workpieces [32]. In MRAFF and R-MRAFF processes, only non-ferromagnetic workpieces could be finished. In MRH process, the electromagnets are positioned inside the cylindrical workpiece. When the ferromagnetic workpiece is finished by MRH process, the magnetic flux is higher on the tool's surface and lower at the workpiece peripherals, as represented in Figure 5.7a [32]. Further, the retained MR polishing fluid on tool's surface due to magnetic field can be used for attaining the superior surface quality.

Initially, the workpiece consists of roughness peaks (Figure 5.7b). The abrasive particles travel to and fro over the internal workpiece surface as per the MRH tool movement. As a result, the surface irregularities are removed and a fine surface finish is obtained, as shown in Figure 5.7c.

For theoretical calculation, the right-angled triangle ΔABC is made from a region ABCD, as shown in Figure 5.8 [33].

The angle (θ) is formed by the arc of the tool surface to vertical tool axis and can be calculated as [39]:

$$\frac{\theta}{2} = \sin^{-1}\frac{w/2}{R_1} \tag{5.8}$$

where R_1 is the radius of the tool outer surface and w is the width of the tool core.

Within the workpiece gap, the area (A) of MR polishing fluid can be calculated as [39]:

$$A = \frac{\theta}{360}\pi\left(R_1^{\,2} - R_2^{\,2}\right) \tag{5.9}$$

where R_2 is the internal cylindrical workpiece radius.

The volume of MR polishing fluid within the working gap can be calculated as [29]:

$$V_{MR\ polishing\ fluid} = A \times h \tag{5.10}$$

where h is the MR polishing fluid height.

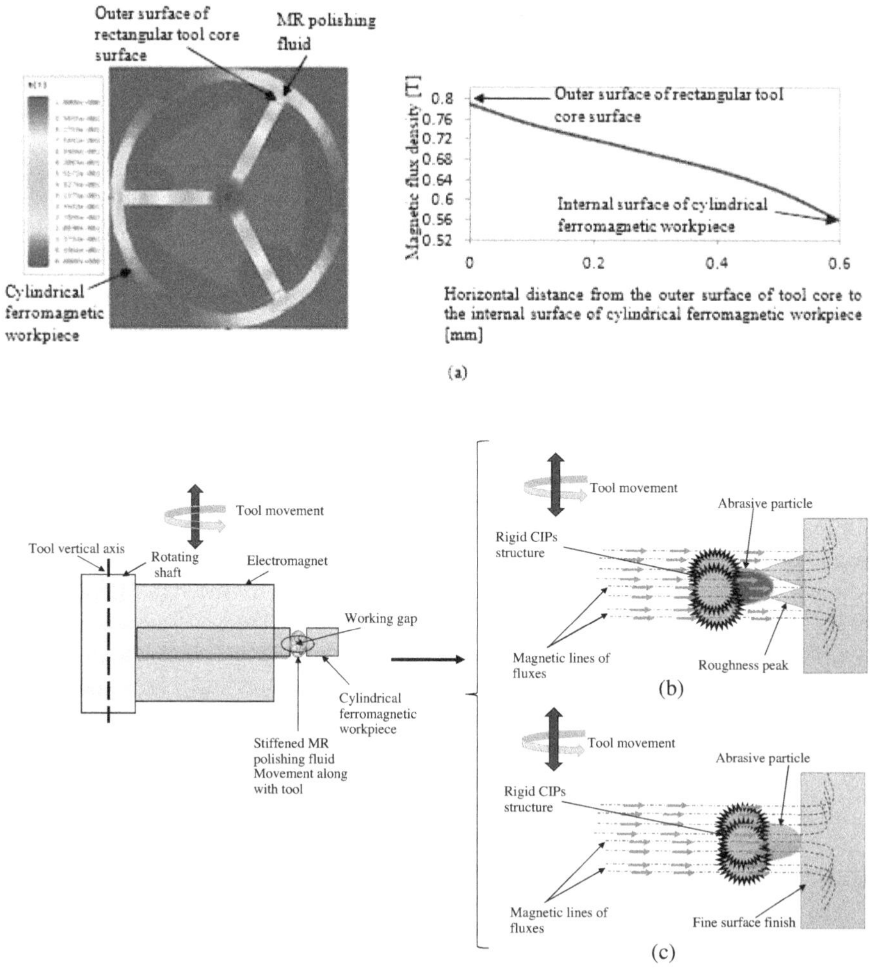

FIGURE 5.7 (a) Magnetostatic simulation with its 2-D field plot among the working gap of MRH tool during internal surface finishing of cylindrical ferromagnetic specimen, and material removal mechanism of MRH process during internal cylindrical surface of ferromagnetic specimen having (b) initial surface and (c) final surface [32].

The magnetic force (F_m) induced by the iron particles as a result of magnetic gradient can be calculated as [39]:

$$F_m = m \times \chi_m \times \mu_0 \times H \times \nabla H \tag{5.11}$$

Bedi and Singh related the capability of two dissimilar designs of MRH tool and found that rectangular-shaped tool provided a uniform magnetic gradient than the I-shaped tool [32]. Further, the surface characteristics of initially honed cast iron cylinder liner were also improved by MRH process [39].

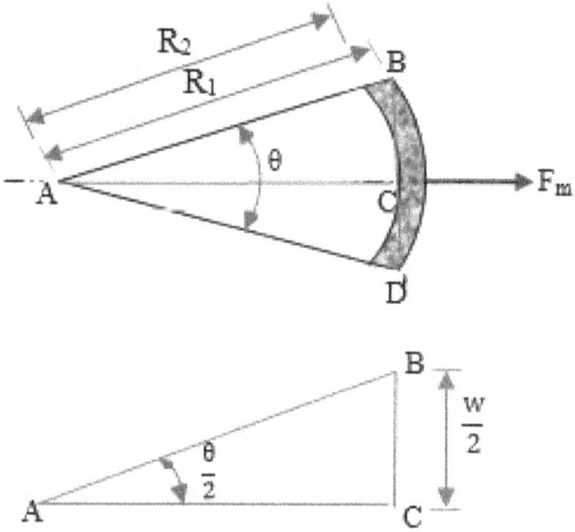

FIGURE 5.8 Line diagram created by means of MR polishing fluid region during MRH process [39].

5.2.5 Magnetorheological Honing (MRH) Process for Variable Internal Diameters

The schematic (front and top views) of permanent magnet-based MRH tool is given in Figure 5.9a [29]. MRH consisted of a conventional honing tool in which abrasive sticks are replaced by the permanent magnetic sticks. These sticks are mounted such that it can easily be attuned as per the internal diameter of the specimen. The MR polishing fluid is inserted within the working gap, as shown in Figure 5.9b. Due to the magnetic gradient, the MR polishing retains at the magnet surface and becomes stiffer [40,41]. The unwanted material is removed by the combined effect of magnetic (F_m) and axial forces (F_a), as shown in Figure 5.9c. The magnetostatic 2-D field report within the working gap of permanent magnet MRH tool is shown in Figure 5.9d. It is shown that the magnetic gradient is higher at permanent magnet and lower at the ferromagnetic cylindrical surface, which means that the present tool is fit for finishing of ferromagnetic internal cylindrical surfaces. This tool is probably not fit for finishing of a hard material because it needs high magnetic flux, which may not be feasible with the permanent magnet.

Grover and Singh evaluated the capability of two different designs of permanent magnet tool and found that the curved-type magnetic sticks provided a better magnetic gradient than the flat-type magnetic sticks [41]. Consequently, it is reported that for getting micro- and nano-finishing over the ferrous and non-ferrous specimens, MR-based finishing technique plays a crucial role.

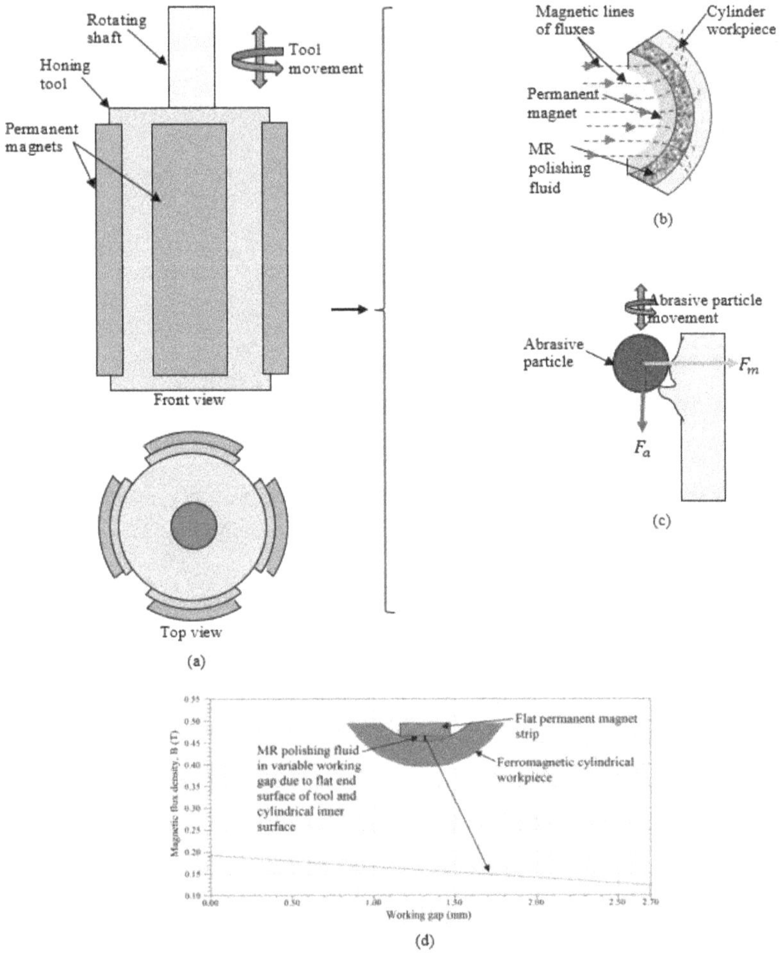

FIGURE 5.9 (a) Schematic of permanent magnet-based MRH tool, (b) MR polishing fluid inserted among the tool surface and specimen surface, (c) different forces acting during material removal mechanism, and (d) 2-D field report within the working gap of permanent magnet MRH tool along with cylindrical ferromagnetic workpiece [29].

5.3 CONCLUSION AND BRIEF DISCUSSION

This study is oriented on conventional and hybrid machining processes to achieve high-quality surface finishing. The machining processes mainly considered the MR finishing processes. Some of the significant conclusions are summarized as follows:

- The conventional finishing processes use solid abrasive stones/sticks for finishing the product's surface. The surface damages are produced over the specimen's surface owing to the shortage in controlling the finishing forces.

- Nonconventional finishing processes which use fine abrasive particles have been adopted to enhance the surface characteristics of the component's surface. However, magnetic-assisted nonconventional finishing processes are able to make a mirror-like surface finish.
- In MR finishing processes, the MR polishing fluid behaves as a semirigid in nature under the consequence of magnetic field, which is further used for improved surface finishing of the component surface.
- However, MRH process is suited for internal cylindrical finishing of ferromagnetic materials, which is likely not finished by MRAFF/R-MRAFF processes.
- From the magnetostatic simulation, it is found that the MRH tool provided an upper magnetic gradient on its tool's surface. This is advantageous for MR polishing fluid to get attached to the tool's periphery only.
- MRAFF process has been applied to finish the internal cylindrical specimens that are non-ferromagnetic in nature. These specimens are made up of copper and aluminum.
- In case of BEMRF process, the three-dimensional surfaces through BEMRF tool have been finished. At the time of finishing, MR polishing fluid enters from the top side of the tool without applying the magnetic field. As soon as MR polishing fluid comes to the tip of the tool, the magnetic field is provided, due to which MR polishing fluid got stiffened and further finishing occurred. It has wide applications for getting finish flat and 3-D curved surface.
- Further, it has been stated that for finishing the internal surface of ferromagnetic specimen, the MRH process has been used. It is specifically used for finishing the ferromagnetic-based materials.

5.4 FUTURE SCOPE

After considering the trending hybrid MR-based machining/finishing process, still some area of scope is still pending, which needed to be carried out in coming studies. The future investigation and research in the area of hybrid machining and finishing are mentioned as follows: There is a need to achieve nano-finishing for brittle materials such as ceramics, glass, and graphites using MR-based finishing process. Using MR-based process, finishing on the intricate shape is needed to be investigated on ferrous and non-ferrous materials.

REFERENCES

1. A. Sharma, V. Jain and D. Gupta, Characterization of chipping and tool wear during drilling of float glass using rotary ultrasonic machining, *Measurement* 128 (2018), pp. 254–263.
2. A. Sharma and V. Jain, Experimental investigation of cutting temperature during drilling of float glass specimen, in *IOP Conference Series: Materials Science and Engineering*, 2020, Shanghai, China.

3. A. Sharma, V. Jain and D. Gupta, Comparative analysis of chipping mechanics of float glass during rotary ultrasonic drilling and conventional drilling: For multi-shaped tools, *Mach. Sci. Technol.* 23 (2019), pp. 547–568.
4. A. Sharma, V. Jain and D. Gupta, Multi-shaped tool wear study during rotary ultrasonic drilling and conventional drilling for amorphous solid, *Proc. Inst. Mech. Eng. Part E J. Process Mech. Eng.* 233 (2019), pp. 551–560.
5. A. Sharma, A. Babbar, V. Jain and D. Gupta, Enhancement of surface roughness for brittle material during rotary ultrasonic machining, *MATEC Web of Conferences*, 249 (2018), pp. 01006.
6. A. Babbar, A. Kumar, V. Jain and D. Gupta, Enhancement of activated tungsten inert gas (A-TIG) welding using multi-component TiO_2-SiO_2-Al_2O_3 hybrid flux, *Measurement* 148 (2019), pp. 106912.
7. A. Babbar, V. Jain and D. Gupta, Thermogenesis mitigation using ultrasonic actuation during bone grinding: A hybrid approach using CEM43°C and Arrhenius model, *J. Brazilian Soc. Mech. Sci. Eng.* 41 (2019), pp. 401.
8. D. Singh, A. Babbar, V. Jain, D. Gupta, S. Saxena and V. Dwibedi, Synthesis, characterization, and bioactivity investigation of biomimetic biodegradable PLA scaffold fabricated by fused filament fabrication process, *J. Brazilian Soc. Mech. Sci. Eng.* 41 (2019), pp. 121.
9. M. Kumar, A. Babbar, A. Sharma and A.S. Shahi, Effect of post weld thermal aging (PWTA) sensitization on micro-hardness and corrosion behavior of AISI 304 weld joints, *J. Phys. Conf. Ser.* 1240 (2019), pp. 012078.
10. A. Babbar, A. Sharma, V. Jain and A.K. Jain, Rotary ultrasonic milling of C/SiC composites fabricated using chemical vapor infiltration and needling technique, *Mater. Res. Express* 6 (2019), pp. 085607.
11. A. Babbar, V. Jain and D. Gupta, Neurosurgical bone grinding, in *Biomanufacturing*, Springer International Publishing, Editor-Chander Prakash, Cham, 2019, pp. 137–155.
12. V. Grover and A.K. Singh, Analysis of particles in magnetorheological polishing fluid for finishing of ferromagnetic cylindrical workpiece, *Part. Sci. Technol.* 36 (2018), pp. 799–807.
13. A. Chana and A.K. Singh, Magnetorheological nano-finishing of tube extrusion punch for improving its functional applications in press machine, *Int. J. Adv. Manuf. Technol.* 85 (2019), pp. 2179–2187.
14. Z. Alam and S. Jha, Modeling of surface roughness in ball end magnetorheological finishing (BEMRF) process, *Wear* 374–375 (2017), pp. 54–62.
15. S. Kumar and A.K. Singh, Nanofinishing of BK7 glass using a magnetorheological solid rotating core tool, *Appl. Opt.* 57 (2018), pp. 942–951.
16. K.E. Oczoś and T.S. Dzioch, Internal spherical surface grinding, *Wear* 133 (1989), pp. 281–294.
17. C. Schmitt, S. Klein and D. Bähre, An introduction to the vibration analysis for the precision honing of bores, *Proc. Manuf.* 1 (2015), pp. 637–643.
18. F. Iqbal and S. Jha, Experimental investigations into transient roughness reduction in ball-end magneto-rheological finishing process, *Mater. Manuf. Process.* 34 (2019), pp. 1–8.
19. V.K. Jain, Magnetic field assisted abrasive based micro-/nano-finishing, *J. Mater. Process. Technol.* 209 (2009), pp. 6022–6038.
20. R. Baraiya, A. Babbar, V. Jain and D. Gupta, In-situ simultaneous surface finishing using abrasive flow machining via novel fixture, *J. Manuf. Process.* 50 (2020), pp. 266–278.
21. A. Babbar, P. Singh and H.S. Farwaha, Regression model and optimization of magnetic abrasive finishing of flat brass plate, *Indian J. Sci. Technol.* 10 (2017), pp. 1–7.
22. A. Babbar, P. Singh and H.S. Farwaha, Parametric study of magnetic abrasive finishing of UNS C26000 flat brass plate, *Int. J. Adv. Mechatronics Robot.* 9 (2017), pp. 83–89.

23. P.S. Gupte, Y. Wang, W. Miller, G.C. Barber, C. Yao, B. Zhou et al., A study of torn and folded metal (tfm) on honed cylinder bore surfaces, *Tribol. Trans.* 51(6) (2008), pp. 784–789.
24. K.D. Lawrence and B. Ramamoorthy, An accurate and robust method for the honing angle evaluation of cylinder liner surface using machine vision, *Int. J. Adv. Manuf. Technol.* 55 (2011), pp. 611–621.
25. S.O.A. El-Helieby and G.W. Rowe, Grinding cracks and microstructural changes in ground steel surfaces, *Met. Technol.* 8 (1981), pp. 58–66.
26. B. Denkena, T. Ortmaier, M. Ahrens and R. Fischer, Monitoring of grinding wheel defects using recursive estimation, *Int. J. Adv. Manuf. Technol.* 75 (2014), pp. 1005–1015.
27. D. Zhu, Y.B. Zeng, Z.Y. Xu and X.Y. Zhang, Precision machining of small holes by the hybrid process of electrochemical removal and grinding, *CIRP Ann. Manuf. Technol.* 60 (2011), pp. 247–250.
28. J.H. Shaikh and N.K. Jain, Modeling of material removal rate and surface roughness in finishing of bevel gears by electrochemical honing process, *J. Mater. Process. Technol.* 214 (2014), pp. 200–209.
29. V. Grover and A.K. Singh, Improved magnetorheological honing process for nano-finishing of variable cylindrical internal surfaces, *Mater. Manuf. Process.* 33 (2018), pp. 1177–1187.
30. S. Jha and V.K. Jain, Design and development of the magnetorheological abrasive flow finishing (MRAFF) process, *Int. J. Mach. Tools Manuf.* 44 (2004), pp. 1019–1029.
31. T.S. Bedi and A.K. Singh, A new magnetorheological finishing process for ferromagnetic cylindrical honed surfaces, *Mater. Manuf. Process.* 33(11) (2018), pp. 1141–1149.
32. T.S. Bedi and A.K. Singh, Development of magnetorheological fluid-based process for finishing of ferromagnetic cylindrical workpiece, *Mach. Sci. Technol.* 22(1) (2018), pp. 120–146.
33. S. Jha and V.K. Jain, Modeling and simulation of surface roughness in magnetorheological abrasive flow finishing (MRAFF) process, *Wear* 261 (2006), pp. 856–866.
34. S. Jha, V.K. Jain and R. Komanduri, Effect of extrusion pressure and number of finishing cycles on surface roughness in magnetorheological abrasive flow finishing (MRAFF) process, *Int. J. Adv. Manuf. Technol.* 33 (2007), pp. 725–729.
35. M. Das, V.K. Jain and P.S. Ghoshdastidar, The out-of-roundness of the internal surfaces of stainless steel tubes finished by the rotational-magnetorheological abrasive flow finishing process, *Mater. Manuf. Process.* 26 (2011), pp. 1073–1084.
36. M. Das, V.K. Jain and P.S. Ghoshdastidar, Nanofinishing of flat workpieces using rotational-magnetorheological abrasive flow finishing (R-MRAFF) process, *Int. J. Adv. Manuf. Technol.* 62 (2012), pp. 405–420.
37. A. Kumar Singh, S. Jha and P.M. Pandey, Design and development of nanofinishing process for 3D surfaces using ball end MR finishing tool, *Int. J. Mach. Tools Manuf.* 51(2) (2011), pp. 142–151.
38. A.K. Singh, S. Jha and P.M. Pandey, Mechanism of material removal in ball end magnetorheological finishing process, *Wear* 302(1–2) (2013), 1180–1191.
39. S.K. Paswan, T.S. Bedi and A.K. Singh, Modeling and simulation of surface roughness in magnetorheological fluid based honing process, *Wear* 376 (2017), 1207–1221.
40. V. Grover and A.K. Singh, Modelling of surface roughness in a new magnetorheological honing process for internal finishing of cylindrical workpieces, *Int. J. Mech. Sci.* 144 (2018), pp. 679–695.
41. V. Grover and A.K. Singh, Modeling of surface roughness in the magnetorheological cylindrical finishing process, *Proc. Inst. Mech. Eng. Part E J. Process Mech. Eng.* 233 (2019), pp. 104–117.

6 Capabilities of Powder-Mixed EDM using Carbon Nanotubes for Biomedical Application

S. Devgan
Khalsa College of Engineering & Technology

S.S. Sidhu
Beant College of Engineering &Technology

A. Mahajan
Khalsa College of Engineering & Technology

CONTENTS

6.1 INTRODUCTION

The demand for precise fabrication processes in the biomedical domain is increasing due to the exponential rise in total joint replacements or surgeries related to human anatomy (Sheikh et al. 2017). The high precision and geometric accuracy are the primary necessities for the effective functionality of implantations. However, stress shielding, poor adhesion, metallic ion toxicity, high corrosion, and excessive wear are some possible causes of failure in metallic implants (Maleki-Ghaleh et al. 2015). An unavoidable difference in implant and bone modulus leads to a stress shielding effect. Thus, wear debris is created at the implant–bone interface, resulting in implant loosening (Mahajan et al. 2019a, b). So, the low coefficient of friction along with hard and good wear-resistant surface resists the malfunctioning of the healthy bones (Nasab et al. 2010).

Similarly, the implant materials should have anticorrosion properties to reduce the undesirable release of metallic ions in the body (Devgan and Sidhu 2020a). The excessive corrosion on the implant surface generates the regular demolition of a material due to a lack of chemical inertness. However, material degradation leads to the undesirable release of metallic ions in blood, biofluids, and variant systems of the body (Devgan and Sidhu 2019b). So the implants' surface should be designed with suitable biological, physicochemical, and tribological properties.

The long-term sustainability of implants is equally linked to the surface morphological properties such as roughness, surface energy, and porosity (Burstein and Pistorius 1995). However, more roughness and higher porosity lead to an increase in surface free energy, which promotes the cell growth and proliferation (Rupp et al. 2006). Nano-shaped morphology imitates the structures of extracellular matrix (ECM) proteins and osteoblastic cells, resulting in better cellular attachment and its proliferation (Das et al. 2008). Although the nano-roughness influences the cellular activities of matching parts since the ECM proteins or collagen fibrils of human bones are composed of nano-sized organics (Fratzl et al. 2004). Figure 6.1 represents the macro-, micro-, and nano-level topology of the implant–bone interface.

Recently, carbon nanotubes (CNTs) are potentially utilized for biomedical applications due to their unique properties (Dresselhaus et al. 1996). CNTs consist of cylinders of graphite along with carbon atoms in the hexagonal arrangement (Ichkitidze et al. 2016). CNTs exhibit unique properties such as favorable weight-to-strength ratio and high Young's modulus (1 TPa). Thus, CNTs are utilized for enhancing the mechanical properties of the substrate (Wang et al. 2008). CNT also has excellent physical features such as unique wetting behavior, high surface energy, and the ability to imitate dimensional similarities of body proteins that promote the bone growth properties (Hussain et al. 2014). These nanomaterials also have anticorrosion properties for sustaining the chemical inertness of coatings (Al-Jumaili et al. 2017).

The powder mixed-electrical discharge machining (PM-EDM) has capabilities to form a recast layer of desired properties and emerged as a novel technique in the field of surface modification (Sidhu et al. 2014). PM-EDM generates the integral nano-textured, hard wear-resistant layer, which resists the implant from corrosion and promotes bioactivity (Devgan and Sidhu 2020b). However, the layer produced by PM-EDM is of superior surface integrity, which escalates the bone in growth (Mahajan and Sidhu 2019a, b).

Figure 6.2 illustrates the working mechanism of the PM-EDM process. During machining, a series of electrical sparks is produced between the workpiece and the tool electrode, which provides a high-temperature heating zone (Sidhu and Bains 2017). The targeted workpiece, along with a small fraction of electrode, was melted in the presence of dielectric medium (Devgan and Sidhu 2019b). The heat-affected area forms the plasma channel by uniting the partial content of the workpiece with the constituents of dielectric medium (Bhui et al. 2018). This phenomenon leads to material and phase transformation of the recast layer. Thus, it results in a change in surface integrity and chemical composition of the substrate surface (Bains et al. 2016).

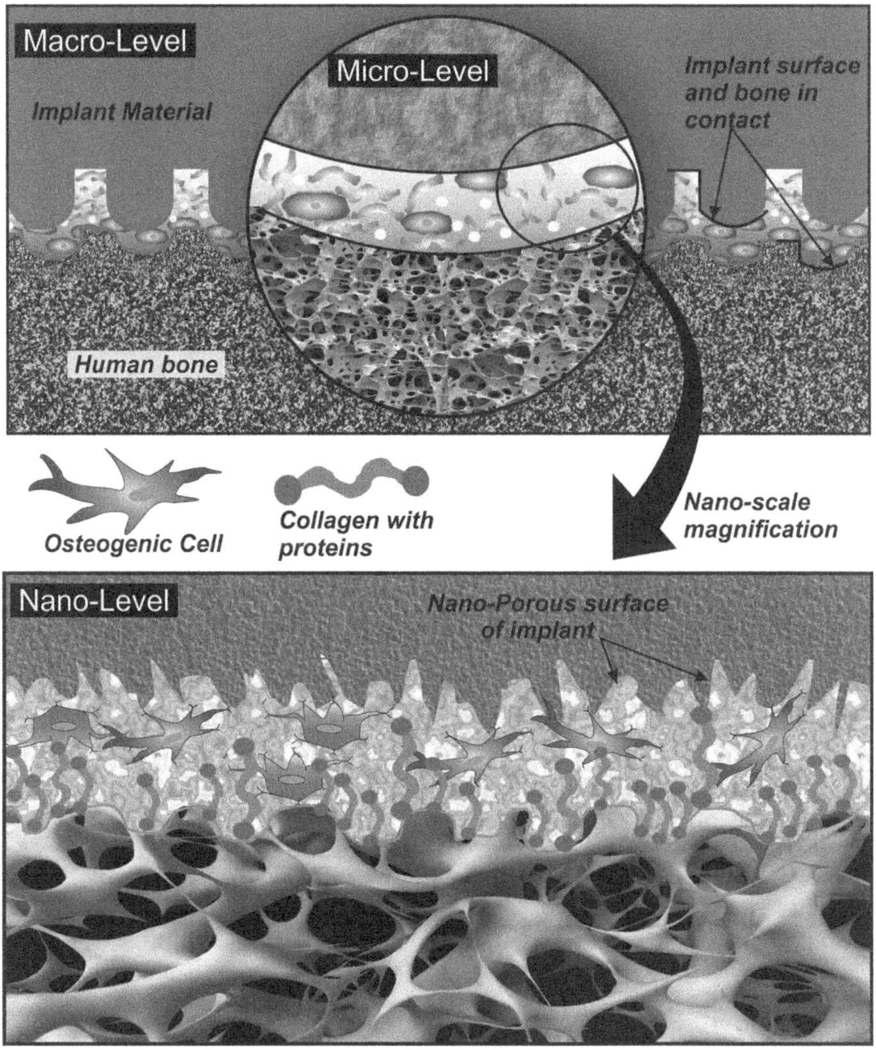

FIGURE 6.1 The macro-, micro-, and nano-level topology of implant–bone interface representing the osteogenic cell growth at nano-textured surface.

Therefore, PM-EDM is employed in this study by utilizing multiwalled carbon nanotubes (MWCNTs) in the dielectric medium. The MWCNTs have extraordinary properties for tailoring the surface integrity of the implant surface. MWCNTs also act as a prominent candidate for increasing the biocompatibility of treated surfaces. In this class, PM-EDM was performed by employing MWCNTs on the β-titanium alloy. This study aimed to scrutinize the PM-EDM processing abilities for attaining the desired surface integrity. The comparison between untreated, deionized water-treated, and MWCNT-treated substrates is scrutinized in terms of bioactivity and surface integrity.

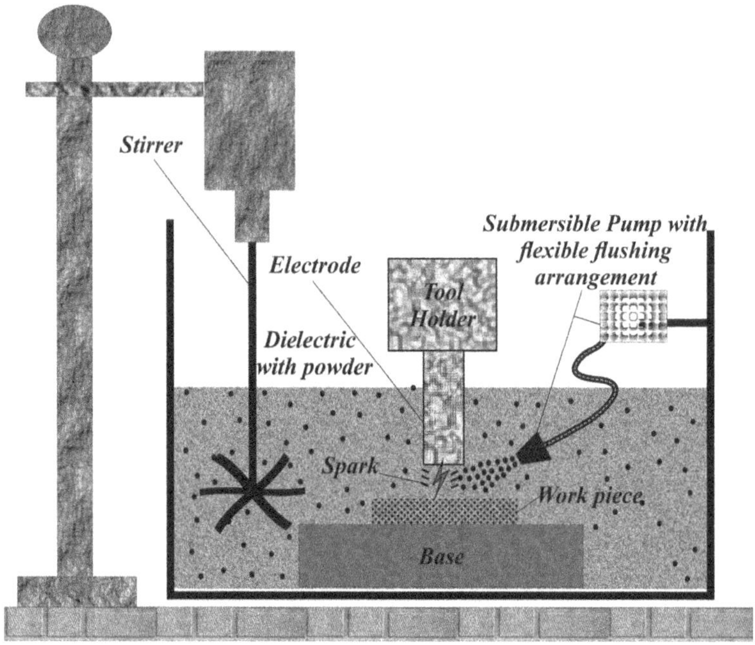

FIGURE 6.2 Schematic arrangement of PM-EDM.

6.2 MATERIAL AND METHOD USED

The β-type titanium alloy (composition (wt.%); Ti: 53%, Nb: 35%, Ta: 7%, Zr: 5%, O: 0.08%) is used for investigation. To investigate the effectiveness of the powder-mixed treatment, experiments were conducted by incorporating two different dielectric mediums, i.e., deionized water and deionized water + MWCNTs. The MWCNTs with a diameter of 10–20 nm and a length of 3–8 µm were procured with 99.9% purity. The dielectric medium has a powder concentration of 5 g/L of deionized water. The graphite (grained ~ 5 µm) was used as a tool electrode material. The experiments were executed on EDM (Model: SZNC-35-5030) at different values of current and pulse on-time and pulse off-time at a predetermined gap of 140 V.

6.2.1 Biocompatibility Testing (In Vitro Hemolysis Test)

The *in vitro* hemolysis test was carried out to scrutinize the biological performance of the treated and untreated samples. Experimentation was executed in triplicate. Autoclaving was carried out at 121°C for 20 min by moist heat sterilization method for the extinction of unwanted entities from the specimens. First, the healthy human blood was extracted in microtubes, and RBCs were separated by centrifuging the blood for 5 min at 2500 rpm at 4°C. The supernatant plasma content and the buffy layer of leukocytes were removed. The dense RBC pellet was obtained by rinsing the supernatant thrice with phosphate-buffered saline (PBS). For the hemolysis, the 1%

RBC suspension was poured on the surface of samples. For the two reference values, Triton XTM-100 was considered a positive control because it can create complete hemolysis. The PBS was utilized as negative control because it has negligible hemolytic properties. Further, the well plate (dish) containing samples were incubated for 1 hour at 37°C. The RBC lysis was carried out in 12-well plates (dishes) in incubator. Hemolytic RBC suspensions from each sample were collected in microcentrifuge tubes. Further, the tubes were centrifuged at 4°C with 2500 rpm for 5 minutes. The absorbance of the supernatant portion was calculated by the spectrophotometric technique at 540 nm. The % hemolysis of each sample was calculated, which is shown in the following equation:

$$\%\text{Hemolysis} = \frac{\text{Abs}}{\text{Abp}} \times 100 \tag{9.1}$$

where Abp is the absorbance value of the positive control sample and Abs is the absorbance value of the tested surface.

6.2.2 Surface Morphological Analysis

The surface morphology is one of the crucial aspects to investigate the processing capabilities of PM-EDM. The surface morphology of the substrates was examined under field emission scanning electron microscopy (FE-SEM) (ZEISS SIGMA VP). The elemental composition of treated substrates was examined using energy-dispersive X-ray spectroscopy (EDX) analysis. The phase compositions were investigated through XRD (XRD-7000 Series, Shimadzu Corporation, Japan) using Cu-Kα X-ray radiation.

6.3 RESULTS AND DISCUSSION

6.3.1 In Vitro Hemolysis Test Results

The biological outcomes of each sample were scrutinized by calculating the absorbance of each sample. The absorbance values were measured thrice for each sample, as listed in Table 6.1. Figure 6.3 presents the percentage of hemolysis of all treated and untreated samples. Triton XTM-100 exhibited the 100% hemolysis, and isotonic PBS was observed with 3.39% ± 0.42% hemolysis. The untreated surface affirmed the % hemolysis of 39.84 ± 1.50. However, MWCNT-treated surface was considered as least hemolytic among all samples. The % hemolysis of the MWCNT-treated sample was calculated as 9.05 ± 3.56, whereas the water-treated surface exhibited a considerable difference to the best value of outcomes. However, % hemolysis of the water-treated sample (i.e., 27.19% ± 2.72%) was twofold higher in comparison with MWCNT-treated substrate (i.e., 9.02%). Similar results were also seen in RBCs' images of all surfaces, as shown in Figure 6.4. MWCNT-treated surface showed the retention of the circular-shaped RBCs, which affirmed that no hemolysis occurs at the surface. The untreated surface had hemolyzed cells and the cells with busted boundaries; however, some clumped cells were seen on the water-treated surface.

TABLE 6.1
Absorbance Values and Percentage Hemolysis of Different Samples

	Attempt (N) = 1		Attempt (N) = 2		Attempt (N) = 3		% Hemolysis
Trial	**Absorbance**	**% Hemolysis**	**Absorbance**	**% Hemolysis**	**Absorbance**	**% Hemolysis**	**Avg ± S.D.**
PBS (negative control)	0.012	3.01	0.013	3.32	0.016	3.84	3.39 ± 0.42
Triton (positive control)	0.398	100	0.391	100	0.416	100	100
Untreated surface	0.153	38.442	0.162	41.43	0.165	39.66	39.846 ± 1.50
Water treated	0.11	27.64	0.116	29.67	0.101	24.28	27.19 ± 2.72
MWCNTs' surface	0.035	8.79	0.022	5.63	0.053	12.74	9.054 ± 3.56

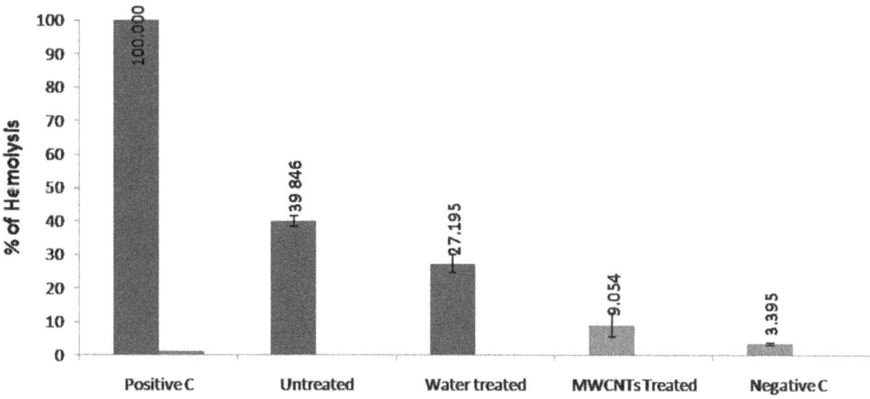

FIGURE 6.3 Hemolysis percentage of different samples.

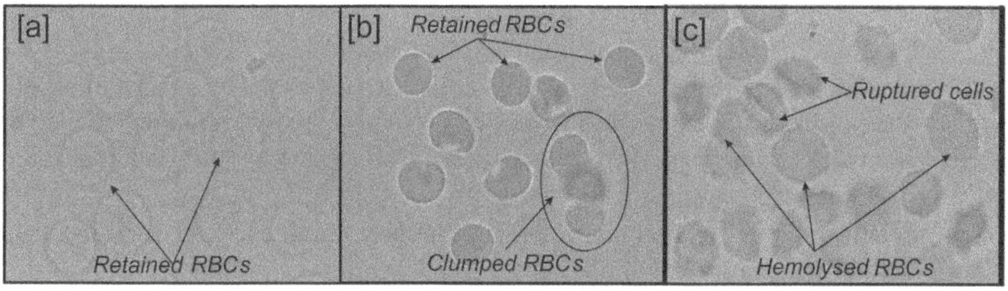

FIGURE 6.4 RBCs' morphology of (a) MWCNT-treated, (b) water-treated, and (c) untreated surfaces.

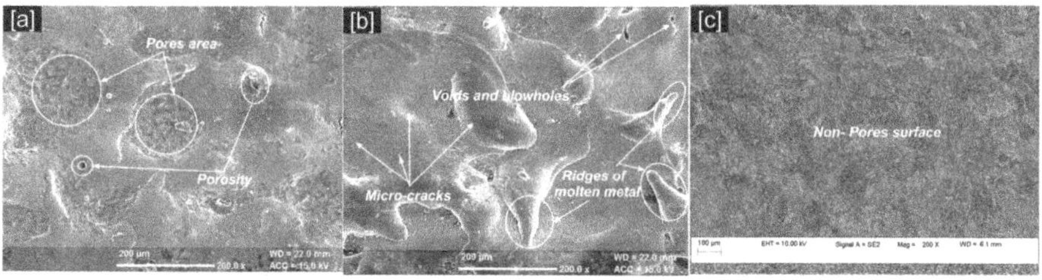

FIGURE 6.5 Surface morphology of (a) MWCNT-treated, (b) water-treated, and (c) untreated substrates.

6.3.2 Surface Morphological Analysis

The surface morphological analysis of the substrates was carried out using FE-SEM. Figure 6.5a–c illustrates the surface morphology of MWCNT-coated, water-treated, and untreated surfaces, respectively. Figure 6.5a shows the microstructure of MWCNT-treated surface, which demonstrates the best surface integrity for bioimplants. The MWCNT surface exhibited a well-patterned structure with

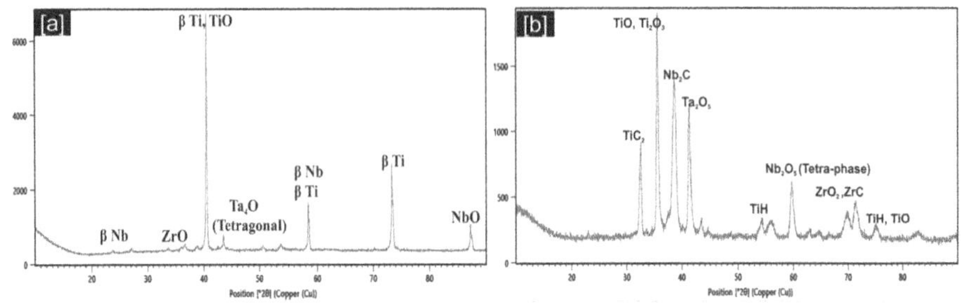

FIGURE 6.6 XRD of (a) untreated and (b) water-treated surfaces.

pores' surface. This was due to the conductive properties of MWCNTs present in the dielectric medium. The MWCNT powder generates a homogeneous heat flux zone that enlarges the surface area of the plasma. The swelling of the plasma channel leads to the proper circulation of spark heat. This phenomenon diminishes the thermal shocks and facilitates the adequate liberation of heat. Thus, it develops the uniform recast layer without any blowholes and microcracks. The surface morphological analysis of water-treated substrate shows some surface irregularities such as voids, pockmarks, and ridges of redeposited material (Figure 6.5b). Also, the water-treated substrate affirms some extent of microcracks and uneven residues of molten metal (Figure 6.5c). Therefore, the PM-EDM generates an efficient surface integrity, which stimulates the cellular activities and demonstrates an excellent tissue–material interaction.

The XRD results of untreated alloy and MWCNT-treated surface are presented in Figure 6.6a–b. The results of the untreated surface affirmed the base metal constituents of alloy, which include β-phases of Ti, Nb, Zr, and Ta. The MWCNT-treated surface (Figure 6.6b) exhibits the compounds of oxides with base constituents such as TiO_2, Ti_2O_3, TaO, ZrO_2, and Nb_2O_5. These nonreactive compounds shaped inert layers on the surface that also endorsed the corrosion resistance of the alloy (Gai et al. 2018). The configuration of oxide phases on the treated substrate is considered as the sympathetic surface topology for increasing the biological performance of the implant material (Jenko et al. 2018). Also, the MWCNT-treated surface exhibited entities of carbon that resulted in the formation of carbides of the base material, i.e., TiC_2, Nb_2C, and ZrC. The compounds of the carbon contributed to the microhardness and wear resistance of the recast layer (Chen et al. 2007). Thus, the carbon compounds, along with oxides, are likely responsible for the formation of the layer having superior tribological properties.

6.4 CONCLUSIONS

This study presents the PM-EDM of β-titanium alloy to achieve the biomimetic surface by incorporating MWCNTs. It explores the capabilities of MWCNTs during machining to produce a biocompatible coating along with desired surface integrity. It concludes that the efficient surface integrity was successfully achieved on the β-titanium alloy as MWCNTs had superior mechanical properties, chemical inertness, and exceptional thermal conductivity. The MWCNTs' substrate exhibited a well-patterned surface with porosity at the nano-scale level, which promotes cellular activities within the body. The biological test concluded that MWCNT-treated surface exhibited superior outcomes of hemolysis (i.e., 9.02%), whereas the untreated sample was considered as most hemolytic. XRD results found that the breakdown of powder entities from dielectric medium and some oxide phases was significantly involved in the formation of the highly inert and biocompatible surface. Also, the MWCNTs' deposition favors the creation of a high-quality wear-resistant surface.

In conclusion, PM-EDM with MWCNTs is a prominent technique to achieve the preferred surface properties that increase the overall performance of the implant.

REFERENCES

Al-Jumaili A, Alancherry S, Bazaka K, Jacob M V. (2017). Review on the antimicrobial properties of carbon nanostructures. *Mater.* 10(9): 1066.

Bains P S, Sidhu S S, Payal H S. (2016). Fabrication and machining of metal matrix composites: A review. *Mater. Manuf. Process.* 31(5): 553–573.

Bhui A S, Singh G, Sidhu S S, Bains P S. (2018). Experimental investigation of optimal ED machining parameters for Ti-6Al-4V biomaterial. *Facta Univ. Ser., Mech. Eng.*, 16(3): 337–345.

Burstein G T, Pistorius P C. (1995). Surface roughness and the metastable pitting of stainless steel in chloride solutions. *Corrosion* 51(5): 380–385.

Chen Y, Zhang T H, Gan C H, Yu G. (2007). Wear studies of hydroxyapatite composite coating reinforced by carbon nanotubes. *Carbon* 45(5): 998–1004.

Das K, Bandyopadhyay A, Bose S (2008). Biocompatibility and in situ growth of TiO_2 nanotubes on Ti using different electrolyte chemistry. *J. Am. Ceram. Soc.* 91(9): 2808–2814.

Devgan S, Sidhu S S. (2019a). Enhancing tribological performance of β-titanium alloy using electrical discharge process. *Surf. Innov.* 8(1–2): 115–126.

Devgan S, Sidhu S S. (2019b). Evolution of surface modification trends in bone related biomaterials: A review. *Mater. Chem. Phys.* 233: 68–78.

Devgan S, Sidhu S S. (2020a). Potential of electrical discharge treatment incorporating MWCNTs to enhance the corrosion performance of the β-titanium alloy. *Appl. Phys. A* 126(3): 1–16.

Devgan S, Sidhu S S. (2020b). Surface modification of β-type titanium with multi-walled CNTs/μ-HAp powder mixed electro discharge treatment process. *Mater. Chem. Phys.* 239: 122005.

Dresselhaus M S, Dresselhaus, G, Eklund P C. (1996). *Science of Fullerenes and Carbon Nanotubes: Their Properties and Applications.* Elsevier, Burlington.

Fratzl P, Gupta H S, Paschalis, E P, Roschger P. (2004). Structure and mechanical quality of the collagen–mineral nano-composite in bone. *J. Mater. Chem.* 14(14): 2115–2123.

Gai X, Bai Y, Li J, Li S, Hou W, Hao Y, Zhang X, Yang R, Misra R D K. (2018). Electrochemical behaviour of passive film formed on the surface of Ti-6Al-4V alloys fabricated by electron beam melting. *Corros. Sci.* 145: 80–89.

Hussain M A, Maqbool A, Khalid F A, Bakhsh N, Hussain A, Rahman J U, Kim M H. (2014). Mechanical properties of CNT reinforced hybrid functionally graded materials for bioimplants. *T. Nonferr. Metal. Soc.* 24: s90–s98.

Ichkitidze L P, Selishchev S V, Gerasimenko A Y, Podgaetsky V M. (2016). Mechanical properties of bulk nanocomposite biomaterial. *Biomed. Eng.* 49(5): 308–311.

Jenko M, Gorenšek M, Godec M, Hodnik M, Batič B Š, Donik Č, Dolinar D. (2018). Surface chemistry and microstructure of metallic biomaterials for hip and knee endoprostheses. *Appl. Surf. Sci.* 427: 584–593.

Mahajan A, Sidhu S S. (2019a). Potential of electrical discharge treatment to enhance the in vitro cytocompatibility and tribological performance of Co–Cr implant. *J. Mater. Res.* 34(16): 2837–2847.

Mahajan A, Sidhu S S. (2019b). In vitro corrosion and hemocompatibility evaluation of electrical discharge treated cobalt–chromium implant. *J. Mater. Res.* 34(8): 1363–1370.

Mahajan A, Sidhu S S, Ablyaz T. (2019a). EDM surface treatment: An enhanced biocompatible interface. In *Biomaterials in Orthopaedics and Bone Regeneration* (pp. 33–40). Editor: Preetkanwal Singh Bains; Beant College of Engineering and Technology, Gurdaspur, India Springer, Singapore.

Mahajan A, Sidhu S S, Devgan S. (2019b). MRR and surface morphological analysis of electrical-discharge-machined Co–Cr alloy. *Emerg. Mater. Res.* 9(1): 1–5.

Maleki-Ghaleh H, Hajizadeh K, Aghaie E, Alamdari S G, Hosseini M G, Fathi M H, Kurzydlowski K J. (2015). Effect of equal channel angular pressing process on the corrosion behavior of type 316L stainless steel in Ringer's solution. *Corrosion* 71(3): 367–375.

Nasab M B, Hassan M R, Sahari B B. (2010). Metallic biomaterials of knee and hip—A review. *Trends Biomater. Artif. Organs* 24(1): 69–82.

Rupp F, Scheideler L, Olshanska N, De Wild M, Wieland M, Geis-Gerstorfer J. (2006). Enhancing surface free energy and hydrophilicity through chemical modification of microstructured titanium implant surfaces. *J. Biomed. Mater. Res. Part A: JBMRGL* 76(2): 323–334.

Sheikh Z, Hamdan N, Ikeda Y, Grynpas M, Ganss B, Glogauer M. (2017). Natural graft tissues and synthetic biomaterials for periodontal and alveolar bone reconstructive applications: A review. *Biomater. Res.* 21(1): 9.

Sidhu S S, Bains P S. (2017). Study of the recast layer of particulate reinforced metal matrix composites machined by EDM. *Mater. Today: Proc.* 4(2): 3243–3251.

Sidhu S S, Batish A, Kumar S. (2014). Study of surface properties in particulate-reinforced metal matrix composites (MMCs) using powder-mixed electrical discharge machining (EDM). *Mater. Manuf. Process.* 29(1): 46–52.

Wang W, Yokoyama A, Liao S, Omori M, Zhu Y, Uo M, Watari F. (2008). Preparation and characteristics of a binderless carbon nanotube monolith and its biocompatibility. *Mater. Sci. Eng. C* 28(7): 1082–1086.

7 Material Removal Rate and Surface Roughness Analyses of ED-Machined SUS-316L with Tungsten Tool

Gurpreet Singh and Sarabjeet Singh Sidhu
Beant College of Engineering and Technology

Tarlochan Singh
Dr. B.R. Ambedkar National Institute of Technology

CONTENTS

7.1 INTRODUCTION

Electric discharge machining, or thermospark, is a versatile technique to process DTM (difficult-to-machine) materials such as titanium, steel, and composites for complex contours with more accuracy and precision compared to other nontraditional methods [1]. In the EDM (electrical discharge machining) process, the removal of material from the workpiece is the phenomenon of evaporation and melting due to a series of discrete sparks between the two terminals (workpiece and electrode) submerged in dielectric medium such as kerosene oil, hydrocarbon oil, and deionized water [2]. The removed material in the form of debris was carried out from the machining area by proper flushing [3]. As EDM is a thermal sparking process, it not only erodes the material from the workpiece surface but also alters the metallurgical elements with the formation of new compounds by the reaction of material used and electrode in the presence of dielectric medium [4]. Today's manufacturing scenario is emergent in the direction of the miniaturization of machine components, devices, etc. which formulate EDM as a dominating process to produce microfeatures in DTM materials such as tungsten carbide, steel, titanium alloys, and MMCs (metal matrix composites) [5]. Therefore, EDM is extensively employed in automobile, aerospace, and die-mold industries for better and precise machining.

EDM was employed on of Al (6351)-SiC and Al (6351)-SiC-B_4C composite materials by Kumar et al. [6] to investigate the surface characteristics of machined samples. They examined the influence of input parameters on the effect of crater formation and heat-affected zone during the process. The rate of metal removal is directly involved in the craters and voids that are present on the machined surface. Sharif et al. [7] machined 316L using EDM and observed that the applied current intensity was the most significant parameter affecting all the output parameters. A similar result was reported by Chaudhury et al. [8] unfolding applied current and pulse duration as the most significant factors influencing the material removal rate (MRR) and tool wear rate (TWR) while using different types of circular electrodes to machine the workpiece material. However, Jaharah et al. [9] concluded that higher values of discharge current lead to rough texture of machined surface. During the machining of H13 tool, they found current intensity as the most significant parameter affecting the MRR as well as the integrity of machined H13 surface. The effect of tool material on the output responses was investigated by Gopalakannan and Senthilvelan [10]. They use graphite, copper, and copper-tungsten tool for 316L and 17-4 PH stainless steel, and found that copper electrode shows a promising potential for higher MRR instead of other electrodes. However, copper-tungsten exhibits high accuracy, better surface finish, and low TWR compared to copper and graphite tool. Das et al. [11] concluded discharge current as the most dominating factor for MRR and surface roughness (SR) of EN31 tool steel. Increased value of applied current along with pulse on-time (P-on) notably affects the erosion rate from the workpiece material.

Apart from this, some researchers also added powder in dielectric to intensify the surface integrity and process rate. Sidhu et al. [12] modified the surface of aluminum-based MMCs in powder-mixed EDM. The output responses were examined in terms of microhardness and improved surface integrity, which were further confirmed by SEM (scanning electron microscopy) and XRD (X-ray diffraction) with the presence of material deposition and compositional analysis. Bains et al. [13] optimized the process parameters for aluminum-based SiC composite using the magnetic field-assisted EDM. The use of hybrid EDM stabilizes the machining process with a decrease in recast-layer thickness and microhardness, and depicts improved MRR and surface integrity of the machined Al-SiC MMCs.

Based upon the literature survey and prominence of EDM, this study is devoted to investigate the performance of tungsten as tool electrode for the EDM of austenitic steel-grade SUS-316L. The output observations were assessed for MRR and SR of the EDMed surface and significant parameters inspected by analysis of variance (ANOVA), based upon their p-values. In addition, regression model is generated with the aid of statistical software Minitab-17 to formulate the relationship between the input factors and a single output response.

7.2 EXPERIMENTAL DETAILS

Thc SUS 316L is eminent material used for heat exchangers, marine industry, and chemical processing equipment, and as metallic biomaterial due to its corrosion resistance performance in acetic acid, hydrochloric acid, and alkaline chloride environment. Herein, the EDM of SUS-316L ($Fe_{69.93}$.$Cr_{16.13}$.$Ni_{10.15}$.$Mo_{2.05}$.$Mn_{1.07}$.$Si_{0.41}$.$P_{0.22}$.$S_{0.017}$.$C_{0.016}$) with a density of 7.99 g/cm^3, a melting range of 1371°C–1399°C, and a thermal conductivity of 21.5 W/m.K at 500°C was performed using tungsten electrode (W-98.3%, ThO_2-1.7%) with a density of 19.25 g/cm^3, a melting point of 3410°C, and a thermal conductivity of 173 W/m.K Taguchi's L_9 orthogonal layout was employed for design of experiments to limit the experimental trials and carried out on ZNC-EDM (OSCARMAX, S645, Figure 7.1). Commercial-grade hydrocarbon oil with a freezing point of 94°C and a specific gravity of 0.763 was utilized as dielectric medium. The prominent machining parameters (peak current, P-on, and

FIGURE 7.1 Pictorial view of experimental setup.

TABLE 7.1
Parametric Conditions for Experimentation

Variable	Units	Symbols	Setup Value
Peak current	A	Ip	20, 24, 28
Pulse on-time	μs	P-on	60, 90, 120
Pulse off-time	μs	P-off	60, 90, 120
Polarity	–	–	Reverse (workpiece negative)
Machining time	min	–	30 (each trial)
Flushing pressure	Kgf/cm^2	–	0.5 (side flushing)

pulse-off-time (P-off)) endorsed by pilot trials varied at three levels, and Table 7.1 lists the parametric conditions for the current experimentation.

In this experimentation, three repetitions of nine experiments were performed on SUS-316L workpiece plate at random order to minimize the noise and errors (*S/N* ratios), and to attain precise output response. The *S/N* ratio is the ratio of magnitude of the desired signal strength to the background noise and represented as:

'*Larger-is-better*': condition for maximizing the response variable in Eq. (7.1), '*Smaller-is-better*': condition for minimizing the response variable in Eq. (7.2).

$$S/N = -10 \times \log 10\left(\text{sum}\left(1/Y^2\right)/n\right) \tag{7.1}$$

$$S/N = -10 \times \log 10\left(\text{sum}\left(Y^2\right)/n\right) \tag{7.2}$$

where y = response and n = repetitions

Wensar made digital weighing balance having a least count of 0.001 g that is used to determine the weight loss of workpiece and electrode after each trial for the calculation of material removal from the surface as per Eq. (7.3).

$$\text{Material removed (mg/min)} = \frac{\text{Initial weight } - \text{ Final weight}}{\text{Machining time}} \times 1000 \tag{7.3}$$

Due to the higher thermal conductivity and melting point of the tungsten, the electrode wear rate was very negligible and hence not considered for the calculation of TWR. The SR of the EDMed surface was calculated as a mean of three readings measured diametrically using 'Mitutoyo-SJ400' roughness tester. Further, the topography of the surface was inspected using SEM (Joel 6610-LV), whereas XRD (PANalytical, X'Pert Pro MPD) examination was used for the phase analysis of machined samples.

7.3 RESULTS AND DISCUSSION

The output responses were investigated in terms of MRR and SR of the machined surface. The observed output values of the responses for each trial are shown in Table 7.2 along with the calculated values for *S/N* ratios. A relative assessment

TABLE 7.2
Experimental Design and Output Response Values

	Process Parameters			Output Response Values							
				MRR (mg/min)			*S/N* Ratios	SR (μm)			*S/N* Ratios
Trial	Ip	P-on	P-off	Rep 1	Rep 2	Rep 3	(dB)	Rep 1	Rep 2	Rep 3	(dB)
1	20	60	60	1.450	1.354	1.339	2.7879	0.372	0.418	0.401	8.0142
2	20	90	90	2.175	2.153	2.172	6.7156	0.678	0.721	0.694	3.1242
3	20	120	120	1.779	1.723	1.846	5.0111	0.827	0.821	0.862	1.5469
4	24	60	90	3.131	3.476	3.367	10.4099	0.892	0.918	0.907	0.8600
5	24	90	120	4.402	4.259	4.161	12.6098	1.145	1.606	1.077	−2.2618
6	24	120	60	4.031	3.976	3.903	11.9735	1.128	1.136	1.249	−1.3808
7	28	60	120	5.347	4.982	5.098	14.2120	1.270	1.613	1.135	−2.6347
8	28	90	60	7.024	6.865	7.195	16.9318	1.130	1.280	1.203	−1.6261
9	28	120	90	5.937	6.278	6.521	15.8917	1.090	1.530	1.318	−2.4437

Note: Rep = repetition.

of MRR and SR was made statistically using ANOVA to validate the influencing parameters affecting the EDM of SUS-316L with tungsten tool.

7.3.1 Parametric Influence on Material Removal Rate

During the EDM process, MRR from the workpiece surface is a primary output response to evaluate the required machining time for product finishing. Figure 7.2 illustrates the mean variation of *S/N* ratios for MRR at different levels of input parameters as per the 'larger-is-better' criterion, where higher *S/N* ratio relates to the maximum MRR. As cleared from the *S/N* ratio plot, a notable rise in MRR witnessed with an increase in the amount of peak current and maximum at 28A. From the ANOVA results, Table 7.3 reveals peak current and P-on as significant parameters affecting the machining rate, based on their p-values. However, P-off was proven as insignificant during the current investigation of ED-machined SUS-316L with tungsten tool under the selected parameters. It can be predicted from the signal-to-noise plot that the peak current of 28A, P-on of 90 μs, and P-off of 90 μs are the optimum machining parameters for the maximum MRR of SUS-316L with tungsten tool.

7.3.2 Parametric Influence on Surface Roughness

In EDM, higher value of discharge current not only improves the MRR but also intensifies the formation of cracks and craters, resulting in roughness on the machined surface. The *S/N* ratio graph (Figure 7.3) is plotted with 'smaller-is-better' condition, and the greater value of *S/N* ratio represents the minimum SR. Peak current

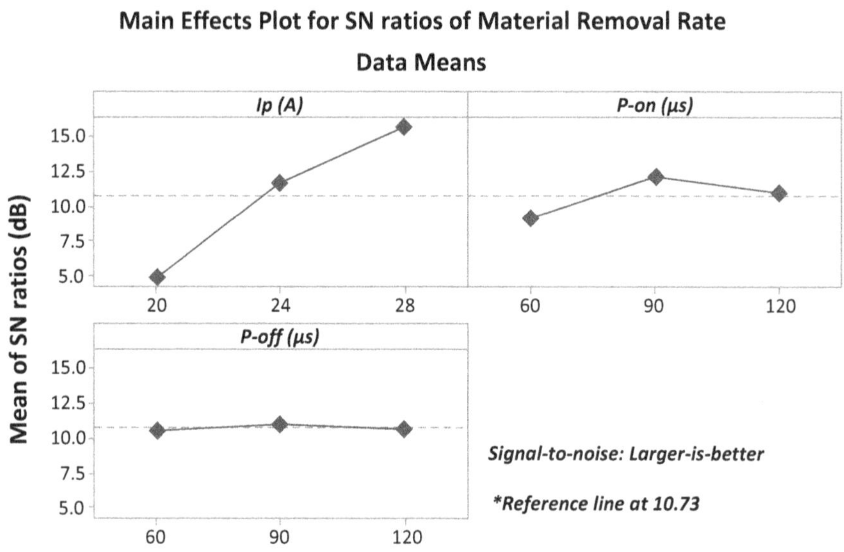

FIGURE 7.2 *S/N* ratio plot for MRR.

TABLE 7.3
Analysis of Variance for SN Ratios of MRR

Source	DF	Seq SS	Adj SS	Adj MS	*F*-Value	*p*-Value
Ip (A)	2	180.223	180.223	90.1117	402.55	0.002*
P-on (μs)	2	13.288	13.288	6.6440	29.68	0.033*
P-off (μs)	2	0.353	0.353	0.1764	0.79	0.559
Residual error	2	0.448	0.448	0.2239		
Total	8	194.312				

* Significant at 95% confidence level.

is the most significant parameter for the roughness of EDMed SUS-316L surface with superior value of 0.397 μm (trial 1). However, the mean *S/N* ratio variation of P-on and P-off is not participated much for the SR, and also, ANOVA possesses p-value > 0.05 (Table 7.4), representing P-on and P-off as insignificant parameters. The optimum parameters for higher surface finish are at the lower level of the input process parameters combination, i.e., peak current of 20A and P-on/P-off duration of 60 μs.

7.3.3 Regression Model and Confirmatory Test

Apart from the ANOVA and optimum values using *S/N* plots, a regression analysis was carried out to correlate the output responses and input variables. The regression

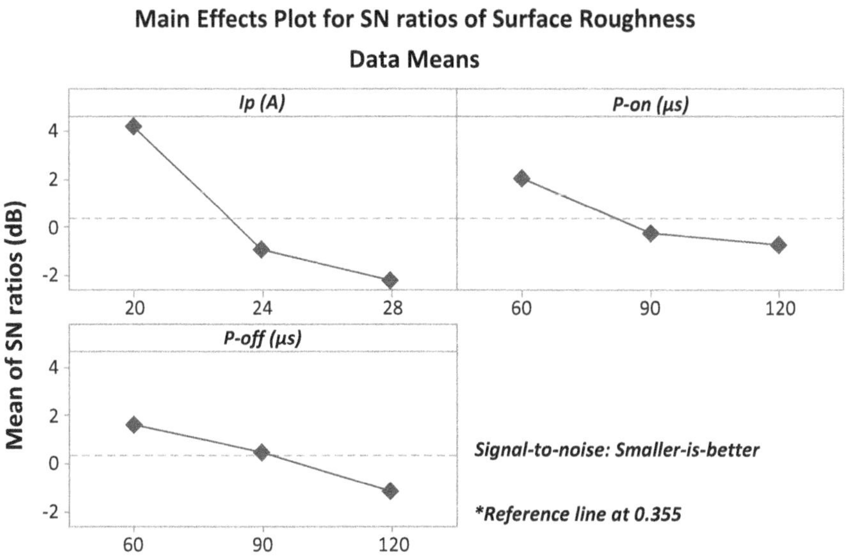

FIGURE 7.3 *S/N* ratio plot for SR.

TABLE 7.4
Analysis of Variance for SN Ratios of SR

Source	DF	Seq SS	Adj SS	Adj MS	*F*-Value	*p*-Value
Ip (A)	2	70.068	70.068	35.034	23.51	0.041*
P-on (μs)	2	13.765	13.765	6.882	4.62	0.178
P-off (μs)	2	11.752	11.752	5.876	3.94	0.202
Residual error	2	2.981	2.981	1.490		
Total	8	98.566				

* Significant at 95% confidence level.

model was created via statistical software Minitab-17, and the equations for MRR (Eq. 7.4) and SR (Eq. 7.5) are as follows:

$$\text{MRR} = -9.65 + 0.5452(\text{Ip}) + 0.01194\ (\text{P-on}) - 0.00656\ (\text{P-off}) \tag{7.4}$$

$$\text{SR} = -1.588 + 0.0802(\text{Ip}) + 0.00377(\text{P-on}) + 0.00378(\text{P-off}) \tag{7.5}$$

where Ip is the peak current (A), P-on is the pulse on-time (μs), and P-off is the pulse off-time (μs).

The regression model was then validated by confirmatory test comparing the predicted value of regression equation to the experimental results. The compared results

TABLE 7.5
Comparison Results of Experimental Values with Regression Model

Exp. No.	Material Removal Rate			Surface Roughness		
	A	B	C (%)	A	B	C (%)
1	1.381	1.576	87.62	0.397	0.469	84.64
2	2.166	1.738	75.37	0.697	0.695	99.71
3	1.782	1.898	93.88	0.836	0.922	90.67
4	3.324	3.560	93.37	0.905	0.903	99.77
5	4.274	3.722	85.16	1.276	1.129	86.97
6	3.970	4.474	88.73	1.171	1.016	84.74
7	5.142	5.544	92.74	1.339	1.337	99.85
8	7.028	6.296	88.37	1.204	1.223	98.44
9	6.245	6.458	96.70	1.312	1.450	90.48

Note: A = mean experimental observations of Rep 1, Rep 2, and Rep 3; B = predicted value of regression equation; C = prediction accuracy (%).

of experimental values with regression equations predicted the model with minimum errors, as shown in Table 7.5.

7.3.4 SEM and XRD Analyses of Machined Surface

The electrical spark generation mechanism between the workpiece and the tool alters the properties of machined surface in terms of its phase transformation and changed surface morphology. It can be noticed from Table 7.2 that higher intensity of the applied current significantly affects the MRR during the machining process. Consequently, thermally influenced surface witnessed nonhomogeneous metallurgical phases in terms of craters, ridges, and cracks. The machined samples were investigated using SEM micrographs and XRD for compositional analysis. In trial 8 (Figure 7.4a), surface exhibiting the maximum material removal illustrates deep craters and greater amount of cracks compared to the sample machined at the lower value of current and P-on for the minimum MRR value in trial 1 (Figure 7.4b).

This can be endorsed to the fact that higher peak current causes continuous sparking at increased P-on causing SR due to the re-solidification of molten metal. Furthermore, it was evident from the previous studies that thermoelectric process changes the chemical composition of the machined surface [14, 15]. XRD analysis (Figure 7.5) validates the presence of newly formed intermetallic compounds such as iron molybdenum ($Fe_{1.92}.Mo_{0.08}$), chromium nickel ($Cr_3.Ni_1$), chromium gallium ($Cr_3.Ga_1$), and molybdenum nickel ($Mo_1.Ni_4$) as a resultant of reaction between substrate elements in the presence of hydrocarbon oil. It was also observed that high value of current and P-on exhibits more crowded surface compared to surface machined at the lower intensity of current. Apart from the listed elements, EDM of

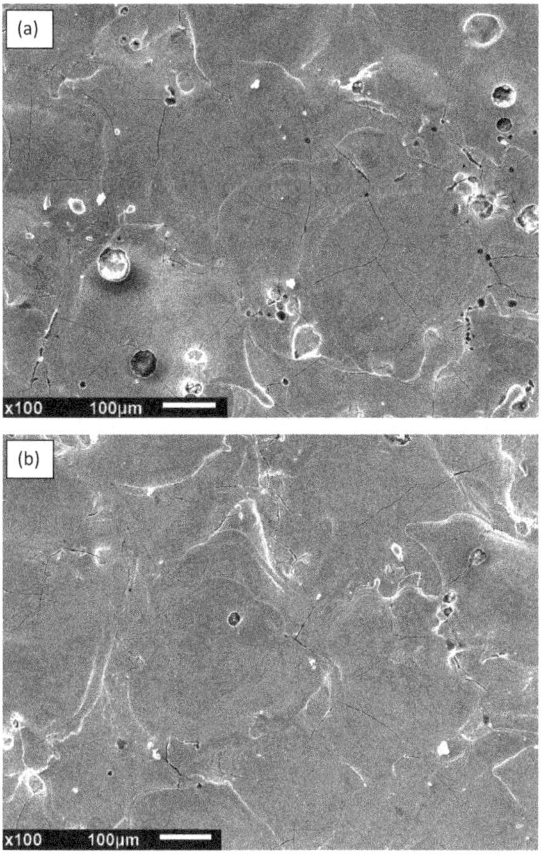

FIGURE 7.4 SEM micrograph of (a) surface with maximum MRR (trial 8) and (b) surface with minimum MRR (trial 1).

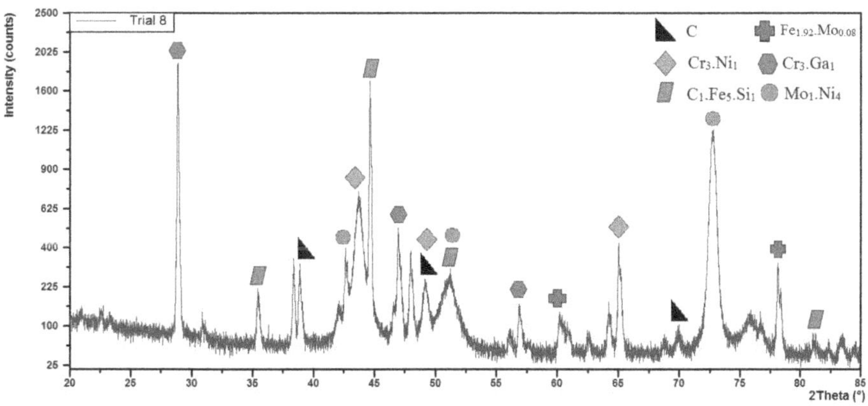

FIGURE 7.5 XRD spectra of ED-machined surface (trial 8).

316L in hydrocarbon oil also depicts the existence of carbon (C) and iron silicide carbide ($C_1.Fe_5.Si_1$) on the machined surface.

7.4 CONCLUSIONS

In this work, the surface of SUS-316L was processed by EDM with tungsten electrode to investigate the machining performance and analyzed the output responses in terms of MRR and SR. The observations drawn from the experimental study are as follows:

- The momentous factors for achieving the maximum MRR were peak current (*p*-value: 0.002), followed by P-on (*p*-value: 0.033). The superior MRR of 7.028 mg/min witnessed at the parameter settings of peak current of 28A, P-on of 90 µs, and P-off of 60 µs.
- Lower values of peak current (20A), P-on (60 µs), and P-off (60 µs) illustrate better surface finish of 0.397 µm as the mean of three repetitions.
- SEM analysis exhibits greater amount of ridges, craters, and cracks at higher value of current (28A) compared to the surface machined at the lower value of the applied current, i.e., 20A.
- Numerous intermetallic compounds and carbide were formed during EDM in the presence of hydrocarbon oil, confirming the changed elemental composition inspected by XRD.

NOMENCLATURE

EDM	Electrical Discharge Machining
DTM	Difficult-to-machine
MMCs	Metal matrix composites
MRR	Material removal rate
TWR	Tool wear rate
SR	Surface roughness
P-on	Pulse on-time
P-off	Pulse off-time
µs	microseconds
µm	micrometer
***S/N* ratio**	Signal-to-noise ratio
dB	decibel
ANOVA	Analysis of variance
SEM	Scanning electron microscopy
XRD	X-ray diffraction

REFERENCES

1. Singh G, Bhui AS, Singh M (2019) Analysis of electrical discharge machining parameters for H13 steel using OVAT technique. *Engineering Research Express* 1:025031.
2. Janmanee P, Muttamara A (2012) Surface modification of tungsten carbide by electrical discharge coating (EDC) using a titanium powder suspension. *Applied Surface Science* 258:7255–7265.

3. Beri N, Maheshwari S, Sharma C, Kumar A (2014) Surface quality modification using powder metallurgy processed CuW electrode during electric discharge machining of Inconel 718. *Procedia Materials Science* 5:2629–2634.
4. Lamichhane Y, Singh G, Bhui AS, Mukhiya P (2019) Surface modification of 316L SS with HAp nano-powder using PMEDM for enhanced biocompatibility. *Materials Today Proceedings* 15:336–343.
5. Bains PS, Sidhu SS, Payal HS (2016) Study of magnetic field assisted ED machining of metal matrix composites. *Materials and Manufacturing Processes* 31:1889–1894.
6. Kumar SS, Uthayakumar M, Kumaran ST, Varol T, Canakci A (2018) Investigating the surface integrity of aluminium based composites machined by EDM. *Defence Technology*. doi:10.1016/j.dt.2018.08.011.
7. Sharif S, Safiei W, Mansor AF, Isa MHM, Saad RM (2015) Experimental study of electrical discharge machine on stainless steel 316L using design of experiment. *Procedia Manufacturing* 2:147–152.
8. Chaudhury P, Samantaray S, Sahu S (2017) Multi response optimization of powder additive mixed electrical discharge machining by Taguchi analysis. *Materials Today: Proceedings* 4:2231–2241.
9. Jaharah AG, Liang CG, Wahid SZ, Rahman MNA, Hassan CHC (2008) Performance of copper electrode in EDM of AISI H13 harden steel. *International Journal of Mechanical and Materials Engineering* 3:25–29.
10. Gopalakannan S, Senthilvelan T (2012) Effect of electrode materials on electric discharge machining of 316L and 17-4 PH stainless steels. *Journal of Minerals and Materials Characterization and Engineering* 11:685–690.
11. Das MK, Kumar K, Barman TK, Sahoo P (2014) Application of artificial bee colony algorithm for optimization of MRR and surface roughness in EDM of EN31 tool steel. *Procedia Material Science* 6:741–751.
12. Sidhu SS, Batish A, Kumar S (2014) Study of surface properties in particulate-reinforced metal matrix composites (MMCs) using powder-mixed electrical discharge machining (EDM). *Materials and Manufacturing Processes* 29:46–52.
13. Bains PS, Sidhu SS, Payal HS, Kaur S (2018) Magnetic field influence on surface modifications in powder mixed EDM. *Silicon* 11:415–423.
14. Singh G, Sidhu SS, Bains PS, Bhui AS (2019) Improving microhardness and wear resistance of 316L by TiO_2 powder mixed electro-discharge treatment. *Materials Research Express* 6:086501.
15. Bhui AS, Singh G, Sidhu SS, Bains PS (2018) Experimental investigation of optimal ED machining parameters for Ti-6Al-4V biomaterial. *Facta Universitatis-Series Mechanical Engineering* 16:337–345.

8 Abrasive Jet Machining
Overview and Scope

Kamaljit Singh Boparai
GZSCCET, MRS Punjab Technical University

Jasgurpreet Singh Chohan
Chandigarh University

CONTENTS

8.1 INTRODUCTION

Since the advent of hybrid composites and hard-to-cut materials, a major shift has been experienced in technology adopted for machining of these high-strength materials. Moreover, there is a need of hour to continuously improve manufacturing processes to meet the demand of highly precise and miniature components for biomedical, optical, and micro-electromechanical systems. The challenge faced by manufacturing industry is to fabricate intricate parts at minimum cost with high surface integrity (Nguyen and Wang, 2019). There are certain conditions where excessive tool wear is a major disadvantage while machining extremely hard materials. In many cases, tool wear and excessive heat generation increase cutting forces inducing defects and seriously deteriorate the surface integrity of parts (El-Hofy et al., 2018).

8.1.1 Abrasive Jet Machining

Although abrasive jet machining (AJM) process is similar to sand blasting, but it differs in the sense that the AJM requires smaller diameter abrasives and has a more finally controlled delivery system. The workpiece material is removed by mechanical abrasion of the high-velocity abrasive particles and is categorized as blast finishing (Wakuda et al., 2003). The schematic of AJM is shown in Figure 8.1. The compressed gas supply (nitrogen, carbon dioxide, or air) passes through a pipe with a high velocity (150–300 m/s) and a flow rate generally maintained at 28 L/min (Shafagh and Papini, 2020). The filtration of gas supply is carried out by filter and dehumidifier. The pressure of the gas is measured by pressure gauge installed after filter, and it is maintained in the range of 2–8 kg/cm^2. The regulated gas supply moves to the mixing chamber. The mixing chamber contains abrasives (such as SiC, Al_2O_3), and the particle size of these abrasives is in the range of 25–30 μm, and has a flow rate 3–25 g/min. Synthetic diamond, SiC and Al_2O_3 powders are generally used for cleaning, cutting, and deburring. In some cases, magnesium carbonate is recommended for superfinishing and itching. Sometimes, sodium carbonate is used for the fine cleaning and cutting of soft materials. Kim et al. (1997) reported the use of pure iron power (80–960 μm) in magnetic AJM. The commercial-grade powder is not suitable for this process because their particle size is not well defined and may contain silica dust, which may cause health hazard. The machining rate of reused

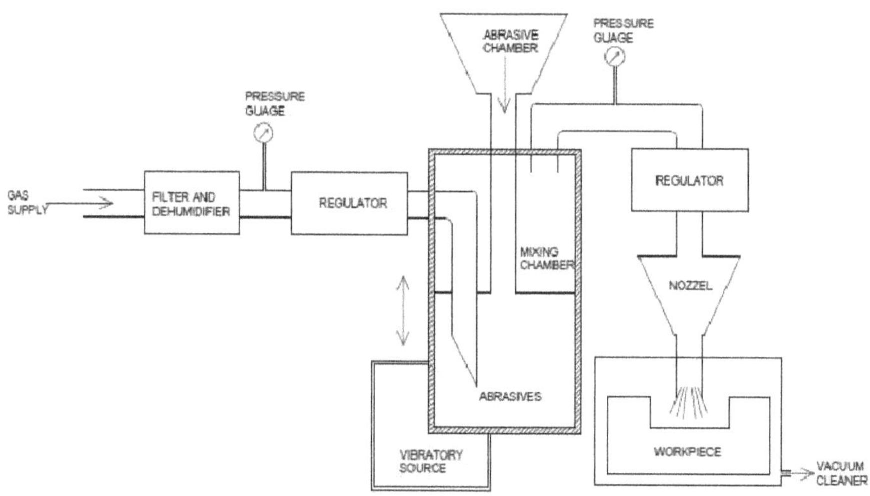

FIGURE 8.1 Schematic of AJM process.

abrasives is declined due to the inclusion of dust and heat. The vibratory source provides the vibration of 50 Hz, and for best performance, the abrasive feed rate is controlled by the amplitude of the vibration in the mixing chamber. The mixture of gas and abrasive from the mixing chamber passes through nozzle. Generally, the material of the nozzle is tungsten carbide or sapphire, and its diameter is in the range of 0.3–2.5 mm. The exit velocity of the nozzle is 150–300 m/s. The jet from the nozzle imping on the workpiece (mounted on machine bed) starts the material removal process by erosion. For ductile materials, the erosion may be caused by various mechanisms such as plastic deformation, fracture, deformation wear (crack and spalling), rupturing, and platelet mechanism. In case of brittle material, it may be happened due to crack formation and crushing (Wakuda et al., 2003; Balasubramaniam et al., 2002). The vacuum cleaner is used for the collection of swarf. The critical input parameters and their respected range are shown in Table 8.1.

The intricate shapes can be produced by masking the surface area of workpiece. As the vibration from the mixing chamber starts, jet starts coming out from the nozzle continuously and hammers the workpiece. Basically, the abrasive particles coming out from the nozzle follow the parallel path for a short distance and afterward follow the narrow cone-shaped path. The abrasive particles hit the workpiece to get the desired shape or cavity. The material removal rate depends on the condition of fluid flow and the wear rate mechanism The larger particle size produces greater material removal rate with poor surface finish, and the smaller particle size produces lower material removal rate with better surface finish. According to 'Finni', the volume of material removed (Q) is calculated as follows:

$$Q = \frac{Cf\,(\theta)MV^{n}}{\sigma}$$

where C and n are the constants, θ is the impingement angle, V is the impacting velocity of particles, and σ is the minimum flow stress of the workpiece.

TABLE 8.1
Critical Input Parameters and Their Respected Range

S. No	Input Parameters	Range
1.	Nozzle diameter	0.3–2.5 mm
2.	Abrasives	SiC, Al_2O_3, pure iron power, magnesium carbonate, sodium carbonate, synthetic diamond powder
3.	Abrasive particle size	25–30 μm
4.	Abrasive flow rate	3–25 g/min
5.	Compressed gas	Nitrogen, carbon dioxide, or air
6.	Stand-off distance	1–4 mm
7.	Nozzle pressure	0.2–1 MPa
8.	Inclination angle	90° for brittle materials. 20°–30° for soft materials.

8.1.1.1 The Advantages of AJM Process Is as Follows

i. No tool changes are required.
ii. Possible to machine thin materials.
iii. In reactive with workpiece material.
iv. Material utilization is high.
v. Suitable for brittle and heat-sensitive materials such as glass, quartz, sapphire, ceramics, and composites.

8.1.1.2 Limitations of AJM

i. Material removal rate is slow.
ii. Silica dust is a health hazard.
iii. Soft materials such as plastics and rubber cannot be machined.
iv. Stray cutting cannot be avoided.

8.1.1.3 Applications of AJM

i. Use for decoration and texturing of window glasses.
ii. Delicate cleaning of irregular surfaces.
iii. Removal of film.
iv. Micro-deburring of hypodermic needles.
v. Machining intricate shapes.
vi. Deburring cross-holes, slots, and threads in small-precision parts used in aircraft fuel system and hydraulic valves.
vii. Wire cleaning without effecting the conductor.

8.1.2 Abrasive Water Jet Machining

Paper Patents Company in Wisconsin, the United States, first used the jet of high-pressured liquid for diagonal cutting of automatically moving sheets of paper in 1933. In earlier days, the applications of abrasive water jet machining (AWJM) process were limited to softer materials like paper due to lower water pressure. Billie Schwacha, in 1958, filed a patent on the behalf of North American Aviation of a system utilizing ultra-high pressure of liquid to cut hard materials. The system was capable to cut stainless steel at 690 MPa pump pressure, but some limitations like delaminating at high speed lead to changes in technology. Numerous studies have reported the use of pulsating and ultrasonic vibrations to assist the AWJM process for higher performance (Tripathi et al., 2018). Recent comparative studies have recognized AWJM as best alternative for conventional machining processes in yielding higher surface integrity, lower compressive residual stress rate, and no thermal stresses with minimum alteration in microstructure while machining free-form surfaces (Holmberg et al., 2019).

8.1.2.1 Definition of AWJM

AWJM is a type of nonconventional machining process where high-velocity jet mixed with abrasive material is impinged on work material. The working principle is simply based on erosion wear of material when a high-velocity jet water (mixed with abrasives) performs the microcutting of upper surface. The major difference between AWJM and pure water jet machining (WJM) lies in erosion behavior and material removal method. The ductile materials demonstrating plastic deformation are preferably cut by WJM, whereas the hard and brittle materials are machined under AWJM as there is risk of cracking due to high impact. In AWJM of ductile materials, material is primarily notched ploughing and microcutting at low impact angle (Finnie, 1960). Conversely, at higher impact angle, the material removal is dominated by plastic failure of the material at point of contact (Bitter, 1953). The advanced version of this nonconventional process is suspension AWJM where injector mixer is eliminated, and hence, abrasive and water are initially mixed at high pressure.

Although WJM is commercially used to cut softer materials such as wood, paper, food materials, polymers, nonmetals and textiles, harder materials such as composites and metals are machined using AWJM. Nowadays, the commercially available AWJM (Figure 8.2) can virtually cut any material ranging from ceramics, metal matrix composites, ceramic matrix composites, layered composites, reinforced plastics, and hard rocks to thick metal plates. The specifications of aforementioned machine are shown in Table 8.2.

In this extended version of WJM, abrasive particles increase the cutting ability to many folds. Most commonly used abrasive particles are diamond, quartz, garnets, boron carbide (B_4C), boron nitride (HBN), silicon oxide (SiO_2), silicon carbide (SIC), or aluminum oxide (AL_2O_3). In many instances, a mixture of two abrasive types in different proportions or two-phase materials are used as abrasives to achieve the desired objectives (Nguyen and Wang, 2019).

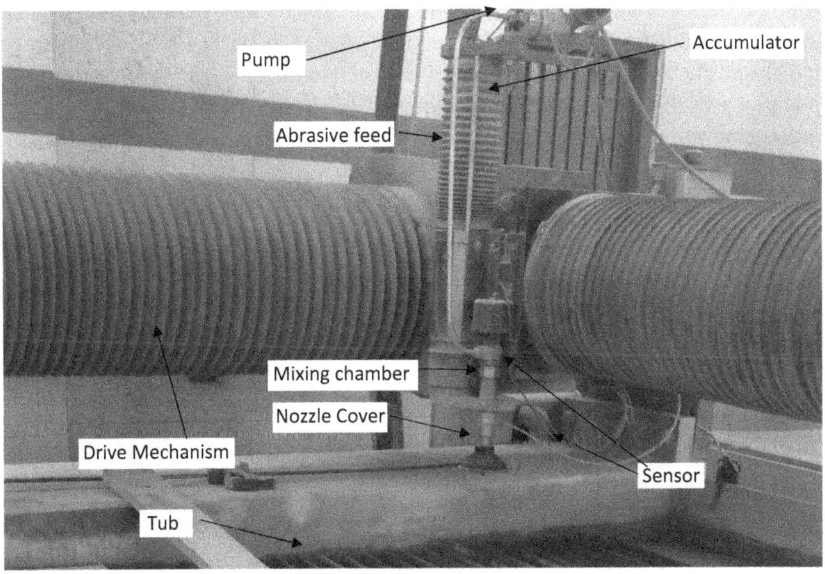

FIGURE 8.2 Components of commercial water abrasive jet machine.

TABLE 8.2
Specifications and Dimensions of Commercially Available Water Jet Machine

S. No.	Parameter	Value
1.	Size (L × B × H)	3937 × 2388 × 2979 mm
2.	Cutting travel (*X*, *Y*, *Z* axis)	2540 × 1397 × 203 mm
3.	Table size	3200 × 1651 mm
4.	Speed	4572 mm/min
5.	Pressure	400 MPa
6.	Linear positional accuracy	±0.025 mm
7.	Repeatability	±0.025 mm
8.	Ballbar circularity	±0.064 mm
9.	Minimum kerf	0.38 mm
10.	Nozzle diameter	0.1778 mm
11.	Energy requirements	3-Phase, 380–480 AC ±10%, 50–60 Hz

8.1.2.2 Applications of AWJM

In addition to cutting, the ultra-high-pressure water jet can be used for pocket milling, drilling, turning, cleaning, paint removal, surgical operations, and peening to remove residual stress. The higher precision and accurate cutting is one of the key strengths of AWJM, as shown in Figure 8.3a, and nozzle is used for cutting (Figure 8.3b).

The process is selectively used for heat-sensitive material due to the absence of heat-affected zone and low-temperature processing. It is best suitable for chemically reactive and hazardous materials which are activated by heat. One of the notable applications of AWJM is dismantling of nuclear power plants due to the absence of heat generation during processing. The complex shapes such as bevels, sharp

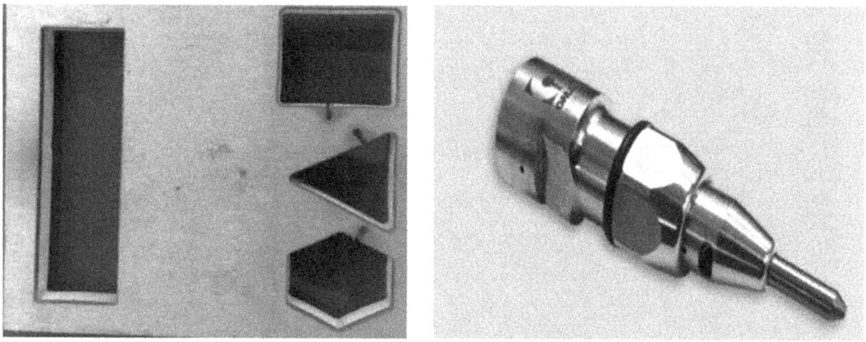

(a) (b)

FIGURE 8.3 (a) Workpiece machined with WAJM and (b) nozzle for WAJM used for machining.

corners, and pierce holes with minimal radii can be easily machined through AWJM attached with CNC head controlled by dedicated software packages.

8.1.2.3 Advantages of AWJM

i. *Absence of heat-affected zone*: No thermal damage as heat is dissipated readily.
ii. *Complex designs*: Make intricate cuts in materials, contours, shapes, and bevels of any angle.
iii. *Cleaner production*: Clean and dust-free process, no metal contamination by cutter, no slag and burrs, absence of smoke and fumes.
iv. *Flexibility*: Able to cut any material (soft of hard).
v. *Environment-friendly*: Minimal pollution or toxic products.
vi. *Precision*: Tolerances of order of ± 0.005 in.
vii. *Easy*: Limited tooling required.
viii. *Low cutting forces*: No cutter induced distortion.

8.1.2.4 Limitations of AWJM

i. Inability to drill flat bottom.
ii. Inability to machine materials that degrade quickly with moisture.
iii. Higher initial cost.
iv. Unable to cut very thick material.
v. High noise levels.

8.1.2.5 Components of AWJM

The hydraulic pump delivers the water from the storage tank to the intensifier at low pressure (approx. 5 bars), as shown in Figure 8.4. Afterward, booster delivers the water to the intensifier by increasing water pressure up to 11 bar.

Hydraulic intensifier is the major component of AWJM setup, which is responsible for significantly increasing the water pressure (3000–4000 bar). The high-pressure water is temporarily stored in accumulator, which acts as a pressure regulator and

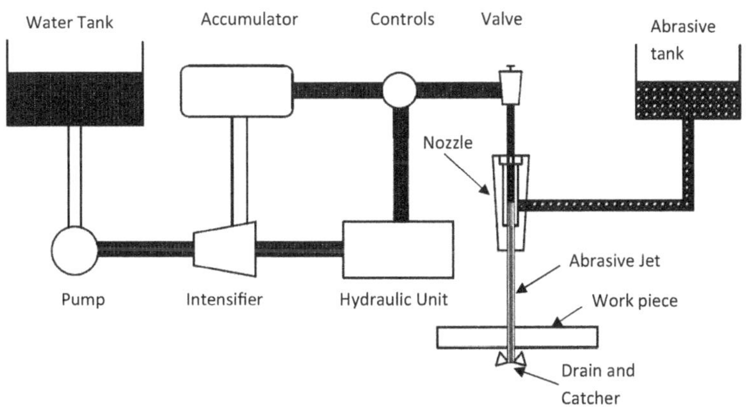

FIGURE 8.4 Schematic of AWJM process.

eliminates fluctuations in water pressure due to varying machining conditions. Afterward, the abrasive particles are mixed with water in the mixing chamber. The direction control and pressure-regulating valves are installed before flow regulator, which facilitates the better control of machining process.

Nozzle is responsible for the conversion of high pressure of water into high-velocity beam of jet. The nozzle is attached to CNC machine tool, which translates in *X* and *Y* direction for microcutting of work material. The ruby or diamond is used to make tip of nozzle to prevent it from erosion. After machining, the water is recirculated through pump before filtering the debris and machined particles through the drain and catcher system.

8.1.2.6 Parameters of AWJM

Modern abrasive water jet equipment operates at extremely high water pressure (200–400 MPa) through intensifier technology. The water is forced to pass through an orifice of very small diameter ranging from 0.2 to 0.4 mm. Hence, the potential energy of water (at high pressure) is converted into kinetic energy leading to high-velocity jet (800–1000 m/s) of water and abrasive mixture. Stabilizers are purposely added to water, which hinders the fragmentation of water jet. Some important parameters of AWJM are water pressure, orifice diameter, abrasive flow rate, stand-off distance, machine impact angle, traverse speed, depth of cut, abrasive grit size, and particle concentration. The optimum settings of various parameters for respective work material as reported by previous case studies are shown in Table 8.3. Moreover, the optimization tool used and the optimum level of parameter are also given in Table 8.3.

The kerf width in case of composite materials is increased with an increase in water pressure and stand-off distance, while it decreases with an increase in feed rate (Kumar et al., 2018). High water pressure, low stand-off distance, and low feed rate, however, yielded straight cuts and better surface finish. Moreover, the impact of pressure is predominant at shorter stand-off distances. The surface roughness is significantly influenced by mesh size of abrasives in AWJM process. In general, the finer abrasive (large mesh size) efficiently removes ductile material and leaves smaller cutting marks. On the other hand, larger-sized abrasives manifest craters on machined surface (Gnanavelbabu et al., 2018). The material removal rate, on the other hand, is higher in case of small orifice diameters and higher water pressure. In addition to this, depth of cut is directly proportional to the mass flow rate and water pressure in AWJM process (Korat and Acharya, 2016). There is a significant impact of impact angle on erosion wear, which was reported by recent studies (Al-Marahleh, 2015) showing the maximum erosion at 90° (for brittle materials) and 20°–30° (for ductile materials).

8.1.3 Hybrid Abrasive Jet Machining

8.1.3.1 Current Status

The AJM has been generally applied for rough machining processes such as deburring and rough/semifinishing. In particular, AJM is required for the machining of structural ceramics, quartz, semiconductors, glass, and electronic devices

TABLE 8.3
Optimum Parameters and Their Levels

Authors	Work Material	Output (Target)	Significant Parameter and Level	Optimization Tool
Sharma et al. (2011)	Coal	Depth of cut (high)	Water pressure (high) = 45 MPa	Taguchi–fuzzy model
Sharma et al. (2011)	Coal	Kerf width (high)	Stand-off distance (high) = 15 mm	Taguchi–fuzzy model
Zain et al. (2011)	AA7075 aluminum alloy	Surface roughness (low)	Transverse speed (low) = 50 mm/min Water pressure (low) = 125 MPa	Genetic algorithm
Kolahan and Khajavi (2011)	6063-T6 aluminum alloy	Depth of cut (high)	Water pressure (high) = 239 MPa Abrasive flow rate (high) = 0.095 kg/min Transverse rate (high) = 125 mm/min	Taguchi
Pawar and Rao (2013)	Stainless steel	Material removal rate (high)	Water pressure (high) = 400 MPa Transverse rate (high) = 150 mm/min	Teaching learning algorithm
Perec (2016)	6082 aluminum alloy	Depth of cut (high)	Water pressure (high) = 28 MPa Transverse rate (high) = 360 mm/min	Taguchi
Tripathi et al. (2018)	Rock	Surface roughness (low)	Transverse rate (low) = 120 mm/min	Full factorial
Kumar et al. (2018)	Aluminum–tungsten carbide composites	Material removal rate (high)	Transverse rate (high) = 360 mm/min Stand-off distance (low) = 2 mm	Response surface methodology
Kumar et al. (2018)	Aluminum–tungsten carbide composites	Surface roughness (low)	Stand-off distance (high) = 6 mm Transverse rate (low) = 190 mm/min	Response surface methodology
Gnanavelbabu et al. (2018)	AA6061-B_4C-hBN MMC	Surface roughness (low)	Particle mesh size (high) = 120 Abrasive flow rate (high) = 440 g/min	Full factorial
El-Hofy et al. (2018)	CFRP Composites	Kerf width (high)	Stand-off distance (high) = 4 mm Water pressure (high) = 350 MPa	Full factorial

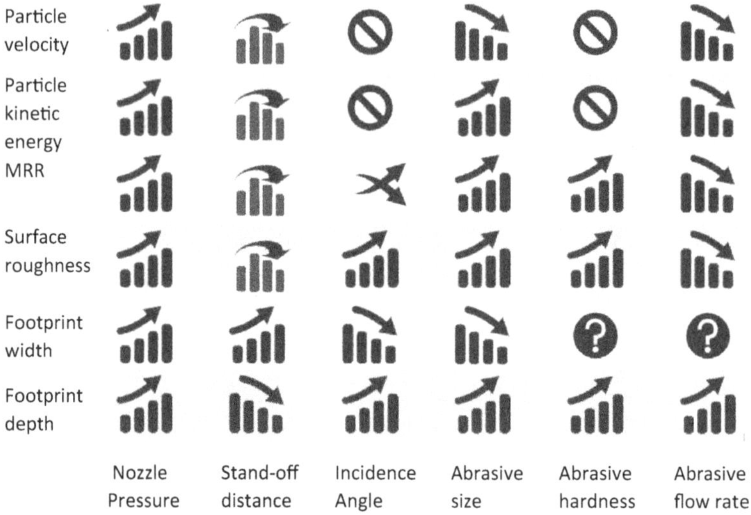

FIGURE 8.5 Effect of input parameters on responses in AJM (Melentiev and Fang, 2018).

(Wakuda et al., 2003; Park et al., 2004). Moreover, rust cleaning from metal surface and surface preparation for welding and plasma spray coatings can also be carried out by AJM (Haldar et al., 2018). Recently, micro-AJM has been expended for the machining of special-purpose part and becomes a useful technique for micromachining of microfluidic and microelectronic devices (Park et al., 2004; Ghobeity et al., 2007). As shown in Figure 8.5, the AJM process has various input parameters such as stand-off distance, mixture ratio, carrier fluid pressure, and grain size that have a significant effect on responses such as material removal rate, surface finish, and penetration rate (Verma and Lal, 1984). Fan et al. (2009) developed the predictive mathematical models to examine erosion rates in microdrilling and microcutting on glasses. AJM is among the unconventional machining processes for various operations such as deburring, cutting, and polishing (Ramachandran and Ramakrishnan, 1993). The geometrical features can be maintained with close tolerances.

8.1.3.2 Opportunities

According to Melentiev and Fang (2018), the classification of AJM is illustrated in Figure 8.6. Kim et al. (1997) reported the use of magnetic AJM for the finishing of parts having circular cross-sectional area (vacuum tubes and sanitary tubes), which are difficult to machine with conventional finishing techniques. Barletta et al. (2007) proposed fluidized bed-assisted abrasive jet machining (FB-AJM) system as an internal finishing process for tubular components. The machining can be carried out on both high-strength and ductile materials (aluminum alloy), and the process will produce comparatively smoother and regular surface.

During the machining of small holes by conventional micro-AJM, the colliding abrasive particles accumulate at the bottom of the hole, which hinders the direct impact of successive abrasives onto the workpiece. Zhang et al. (2005) proposed

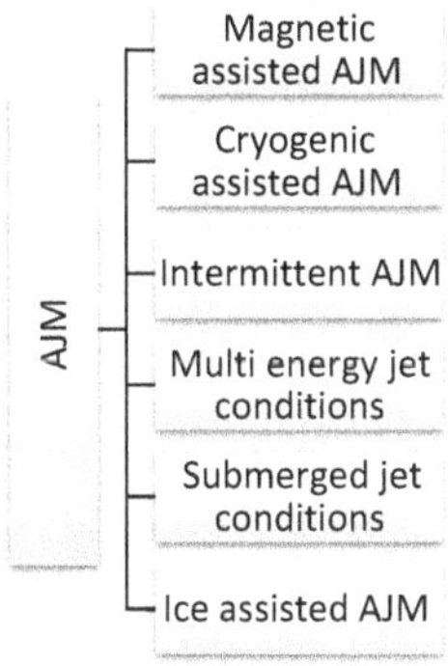

FIGURE 8.6 Classification of AJM.

microabrasive intermittent jet machining (MAIJM) as a solution of the above-said problem. There exists a time lag during which no abrasive is injected into the gas stream so that the continuous flow of gas without abrasives from the nozzle could sweep away accumulated abrasives.

Gradeen et al. (2012) investigated the cryogenic AJM process for the machining of polydimethylsiloxane (PDMS) using aluminum oxide particles in the temperature ranges of –178°C and 17°C, and concluded that the erosion rate of PDMS increased by a factor of more than 10. Kruusing et al. (1999) demonstrated the combined process of AJM and laser beam machining in water for the rapid prototyping of silicon microstructures for fluidic devices. Jafar et al. (2013) reported that the surface finish obtained by micro-AJM can affect fluid flow phenomena such as separation efficiency, electroosmotic mobility, and solute dispersion in microfluidic applications. Fähnle et al. (1998) also highlighted the applications of AJM for locally shaping and polishing of optical surfaces having complex shapes. The limitations of conventional AJM have been eliminated by another new AJM process (Kuriyagawa et al., 2001). In this process, direct etching and patterning of glass can be done without masking.

The machining characteristics (such as material removal rate, electrode wear rate, and surface roughness for a tool steel) of a hybrid process (electrical discharge machining in gas with AJM) have been explored by Lin et al. (2018). They concluded that the proposed hybrid process could enhance the machining efficiency and may fit the requirements of modern manufacturing applications. In order to control the geometric features of parts and to improve the concentration of flow with guiding pressure, a Laval type of nozzle has been recommended for the machining of metals like stainless steel ANSI 316 plate (Vanmore and Dabade, 2018).

Various innovations and iterations (such as intermittent, submerged, thermally assisted, and other jet conditions) in AJM process have been discussed for process improvement, but still it faces some challenges. Further, Wang et al. (2020) experimentally investigated the use of pressurized air wet abrasive jet machining (WAJM) for the edge preparation of cutting tools. Nanda et al. (2020) studied the combined effect of hot silicon carbide abrasive particles with compressed air and used the mixture through the FB-AJM setup for the machining of hard and brittle K-60 alumina ceramic composite material.

8.1.3.3 Challenges

The AJM process is in the transition phase from macro- to micro-scale. Presently, in micromachining resolution, process controlling and erosion prediction are the matters under consideration (Melentiev and Fang, 2018; Gnanavelbabu et al., 2020). For the machining of precise hole with conventional AJM, small-diameter nozzle is required (Kuriyagawa et al., 2001). However, the fine abrasive particles (micro- and nano-scaled) are required for sharp edge formation and high-quality surface finish. Many studies (Zhang et al., 2005; Kuriyagawa et al., 2001) also revealed that with fine abrasive particles, the nozzle clogged and it is difficult to regulate the abrasive flow rate. Generally, in traditional AJM, mineral abrasives are used that are nonenvironment-friendly. Moreover, these abrasives consume nonrenewable resources and cause pollution. Like other machining processes, all AJM processes also have a wear component, i.e., nozzle, which is most susceptible to wear.

8.1.4 Hybrid Abrasive Water Jet Machining

8.1.4.1 Current Status

Many researchers have reported a significant improvement in machining efficiency with an integration of different nonconventional processes with AWJM (Figure 8.7). Molian et al. (2008) combined CO_2 laser cutting technology with AWJM to enhance the cutting of ceramics. The laser produces precise and small heat-affected zone on surface of ceramics, and AWJM facilitates the quenching and results in cracking of material under thermoelastic stresses. The hybrid method improved cutting efficiency and also assisted in washing and cleaning of cracked material. Similarly, hybrid laser/AWJM process was used by Kalyanasundaram et al. (2008) who performed AWJM to study the fracture behavior of alumina along with temperature distribution and thermal stress fields. Analytical model was compared with experimental data to create a map that indicates the impact of individual parameter on thermal shock fracture. The hybridization of CO_2 laser and AWJM, on the one hand, reduces the temperature during material removal, resulting in lower thermal

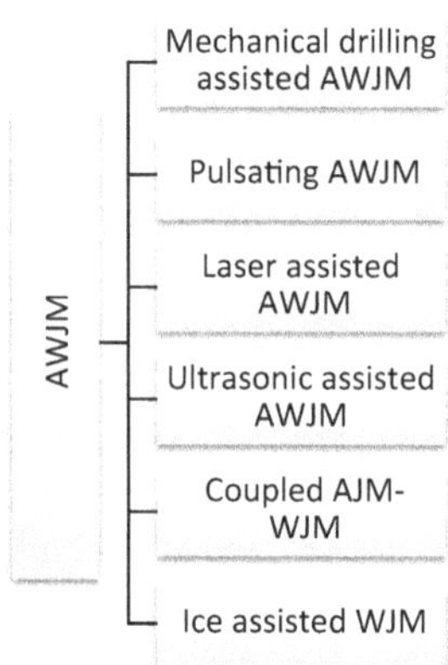

FIGURE 8.7 Classification of AWJM.

damages, and also needs lesser heat energy input that allows higher laser traverse and cutting speed rate.

Ice-jet technology has been added to AWJM where ice particles are added in water jet instead of abrasives. The hybrid process not only reduces cost of production but also reduces environmental damage due to conventional abrasives. Studies reported the use of cryogenic cutting where liquid nitrogen jets are used instead of water to cut softer materials. Many researchers are currently working on the use of advanced cutting fluids to control the thermal and frictional conditions existing at cutting zone (Kovacevic et al., 1997).

Lu et al. (2013) used AWJM to assist the mechanical drilling of hard rocks, and it was concluded that combining two processes increased drilling depth by 63%, and thrust force and torque were reduced by 15% and 20%, respectively. A hard rock-breaking drill was designed having provision of water jet inside which rotates and impinges water jet at same time to improve the cutting efficiency. Kowsari et al. (2016) studied the material removal mechanism in coupled abrasive slurry jet and air jet micromachining processes during machining of aluminum nitride, alumina, and zirconium tin titanate.

Tripathi et al. (2018) utilized pulsating water jet through frequency modulation at 20.20 kHz. It was observed that deeper holes can be drilled with pulsating jet as compared to normal WJM system. Thus, the hybridization of AWJM has been extensively experimented by various researchers yielding positive results in terms of accuracy, cutting speed, surface finish, and manufacturing cost.

Hybrid AJM–WJM process was used by Huang et al. (2013) for upper surface removal of titanium alloy (Ti-6Al-4 V with an alpha case layer) followed by WJM for cleaning of embedded particles and finishing of work surface. It was further observed that grit removal process not only depends on working parameters of normal WJM, but also is primarily decided by embedment behavior of the abrasives. Recently, the combination of electrochemical machining and AJM has been proposed (Gao et al., 2020), as shown in Figure 8.8a. Zhang et al. (2020) conducted experiments on electrochemical abrasive jet processing (ECAJP) for the high-precision machining of Inconel 718. During machining, a mixture of an electrolyte solution and abrasive

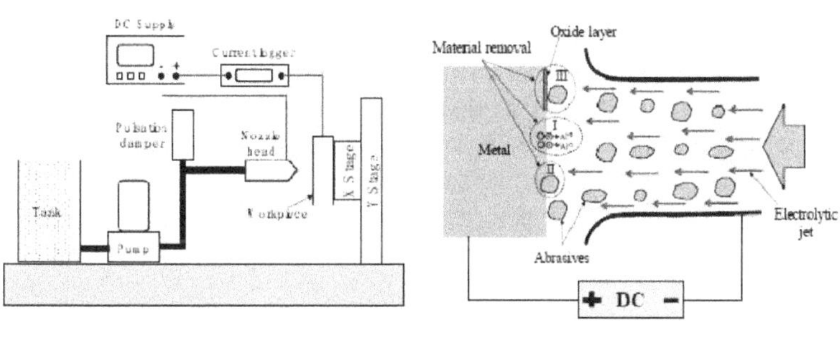

(a) (b)

FIGURE 8.8 (a) Material removal process in ECAJM and (b) schematic of ECAWJM (Gao et al., 2020).

particles was sprayed from the nozzle and impacted the surface of a material while a direct current potential was applied (Figure 8.8b).

8.1.4.2 Challenges

Although AWJM has proven its efficacy for various engineering applications and cutting wide range of materials, there are certain situations where this process has limited scope. The first situation is moisture-sensitive materials that are highly reactive to water and vapors, and thus, it is difficult to machine those materials with this process. During cutting of thick alloys, surface finish is not appropriate and post-processing is required for cleaning and deburring of work material. In the future, the researchers should focus on optimization cutting speed and cutting angle for different materials. The selection of these parameters is highly dependent on mechanical properties of material. In addition to this, future optimization studies must concentrate on the use of advanced optimization tools in case of composite materials having a range of mechanical and chemical properties. Hybridization of AWJMA along with other nonconventional machining techniques is major challenge ahead for researchers as the combined effects are unpredictable and efficacy of coupled system may depend on type of work material.

8.1.5 Recent Advancements in Water Abrasive Jet Technology

Erichsen et al. (2004) patented a technology that controls the abrasive jet system based on real-time data received during machining process. The computer-generated model predicts the output characteristics of work material based on ongoing processing parameters. The system recovers the data automatically, which facilitates the estimation of surface finish, accuracy, and manufacturing time based upon jet speed, diameter, and other operating characteristics. Similarly, Bader et al. (2011) patented a device for generating and controlling high-pressure jet for WAJM system.

Alberts et al. (2008) brought the AJM to one step further by introducing abrasive fluid jet milling technology, which overcomes the disadvantage of overabrasion of work material while using conventional WAJM. The CNC-controlled technology is capable to mill customized pockets in solid work material. The technology facilitates the movement and impinging of jet at desired angle while having three-dimensional motion. This can help to make curvilinear cuts and intricate design on the work material. Another patented product utilizes the combined impact of WJM and electric discharge machining (EDM), as shown in Figure 8.9. The process enables easy cutting of hard-to-cut conductive materials like cemented carbide into desired shape. The work material is mounted on the same bed on which both machines work one after another using retractable apparatus. Initially, the cutting is performed by water jet which is followed by wire-cut EDM process to achieve higher finish and accuracy (Yuzo and Toi, 2009).

Further advancement in WAJM was achieved by attaching real-time tracking device using a number of sensors between a work surface and first plane perpendicular to longitudinal axis of the nozzle. The digital data is transferred to output regarding speed, angle, and accuracy, and also collision detection mechanism ensures damage to costly equipment (Wakefield et al., 2008).

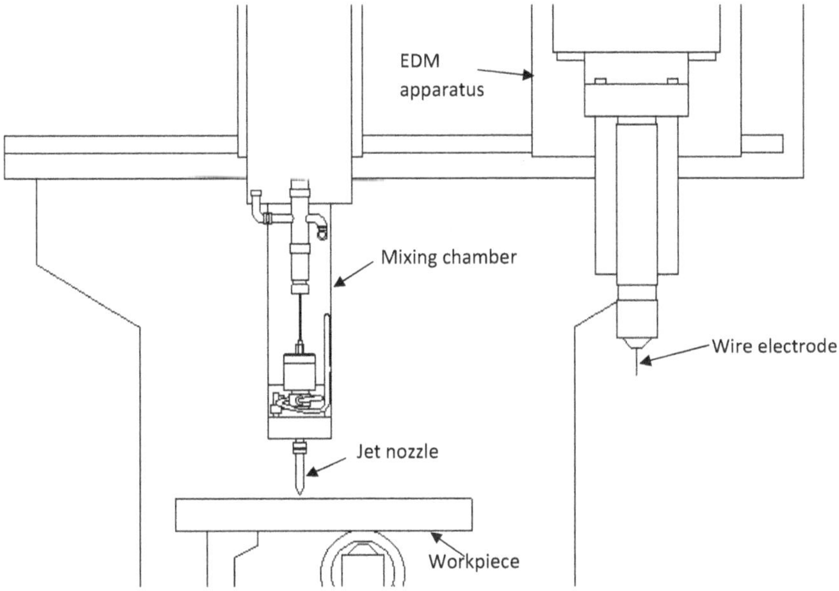

FIGURE 8.9 Schematic of combined EDM and WAJM setup.

8.1.6 Conclusions and Outlook

In the nutshell, AJM and AWJM processes have been successfully implemented for micromachining of various metals and nonmetals, but their applications for the machining of composites and ceramics cannot be overruled. AJM and AWJM are erosion-based manufacturing methods, growing continuously and catering recent and oncoming industrial demands. Presently, among these technologies, the trend of advancements is a shift from the macro- to micro-scale in addition to condition monitoring, hybrid machining, and computer-controlled manufacturing. A variety of hybrid methods have been developed but the reduction of machining spot, precise erosion predictability, and process controlling are current challenges, which require consideration. Finally, the future investigations should be focused on the real-time monitoring, cost optimization, lead time reduction, and precision.

REFERENCES

Alberts, D.G., Cooksey, N., Butler, T.J. and Miles, P.J., Ormond LLC, 2008. CNC abrasive fluid-jet milling. U.S. Patent 7,419,418.

Al-Marahleh, G., 2015. Parameters controlling abrasive water jet technology: Erosion and impact velocity for both ductile and brittle materials. *IOSR Journal of Engineering*, *5*(8), pp. 1–7.

Bader, D.C., Burnham, C.D., Hashish, M.A., Knaupp, M., Mann, R.J., Meyer, A., Pesek, T.A., Sahney, M.K., Stewart, J.M., Vaughan, S.A. and Flow International Corp. 2011. Device for generating and handling a high pressure fluid jet. Patent ES2353267T3.

Balasubramaniam, R., Krishnan, J. and Ramakrishnan, N., 2002. A study on the shape of the surface generated by abrasive jet machining. *Journal of Materials Processing Technology*, *121*(1), pp. 102–106.

Barletta, M., Guarino, S., Rubino, G. and Tagliaferri, V., 2007. Progress in fluidized bed assisted abrasive jet machining (FB-AJM): Internal polishing of aluminium tubes. *International Journal of Machine Tools and Manufacture*, *47*(3–4), pp. 483–495.

Bitter, J.G.A. 1953. A study of erosion phenomenon part I. *Wear*, *6*, pp. 5–21.

El-Hofy, M., Helmy, M.O., Escobar-Palafox, G., Kerrigan, K.M., Scaife, R. and El-Hofy, H., 2018. Abrasive water jet machining of multidirectional CFRP laminates. *Procedia CIRP*, *68*, pp. 535–540.

Erichsen, G.A., Zhou, J., Sahney, M.K. and Knaupp, M. and Flow International Corp, 2004. Method and system for automated software control of waterjet orientation parameters. U.S. Patent 6,766,216.

Fähnle, O.W., Van Brug, H. and Frankena, H.J. 1998. Fluid jet polishing of optical surfaces. *Applied Optics*, *37*(28), pp. 6771–6773.

Fan, J.M., Wang, C.Y. and Wang, J., 2009. Modelling the erosion rate in micro abrasive air jet machining of glasses. *Wear*, *266*(9–10), pp. 968–974.

Finnie, I. 1960. Erosion of surfaces by solid particles. *Wear*, *3*(2), pp. 87–103.

Gao, C., Liu, Z., Qiu, Y. and Zhao, K. 2020. Modelling of geometric features of micro-channel made using abrasive assisted electrochemical jet machining. *International Journal of Electrochemical Science*, 15, pp. 94–108.

Ghobeity, A., Getu, H., Krajac, T., Spelt, J.K. and Papini, M., 2007. Process repeatability in abrasive jet micro-machining. *Journal of Materials Processing Technology*, *190*(1–3), pp. 51–60.

Gnanavelbabu, A., Rajkumar, K. and Saravanan, P., 2018. Investigation on the cutting quality characteristics of abrasive water jet machining of AA6061-B4C-hBN hybrid metal matrix composites. *Materials and Manufacturing Processes*, *33*, pp. 1313–1323.

Gnanavelbabu, A., Surendran, K.S. and Rajkumar, K., 2020. Performance evaluation of abrasive water jet machining on AA6061-B 4 C-HBN hybrid composites using Taguchi methodology. In *Advances in Unconventional Machining and Composites*, Editor: M. S. Shunmugam (pp. 651–660). Springer, Singapore.

Gradeen, A.G., Spelt, J.K. and Papini, M., 2012. Cryogenic abrasive jet machining of polydimethylsiloxane at different temperatures. *Wear*, *274*, pp. 335–344.

Haldar, B., Ghara, T., Ansari, R., Das, S. and Saha, P., 2018. Abrasive jet system and its various applications in abrasive jet machining, erosion testing, shot-peening, and fast cleaning. *Materials Today: Proceedings*, *5*(5), pp. 13061–13068.

Holmberg, J., Berglund, J., Wretland, A. and Beno, T., 2019. Evaluation of surface integrity after high energy machining with EDM, laser beam machining and abrasive water jet machining of alloy 718. *The International Journal of Advanced Manufacturing Technology*, *100*, pp. 1575–1591.

Huang, L., Kinnell, P. and Shipway, P.H., 2013. Parametric effects on grit embedment and surface morphology in an innovative hybrid waterjet cleaning process for alpha case removal from titanium alloys. *Procedia CIRP*, *6*, pp. 594–599.

Jafar, R.H.M., Spelt, J.K. and Papini, M., 2013. Surface roughness and erosion rate of abrasive jet micro-machined channels: Experiments and analytical model. *Wear*, *303*(1–2), pp. 138–145.

Kalyanasundaram, D., Shrotriya, P. and Molian, P., 2009. Obtaining a relationship between process parameters and fracture characteristics for hybrid CO_2 laser/waterjet machining of ceramics. *Journal of Engineering Materials and Technology*, *131*(1), pp. 1–10.

Kim, J.D., Kang, Y.H., Bae, Y.H. and Lee, S.W., 1997. Development of a magnetic abrasive jet machining system for precision internal polishing of circular tubes. *Journal of Materials Processing Technology*, *71*(3), pp. 384–393.

Kolahan, F. and Khajavi, A.H., 2011. Modeling and optimization of abrasive waterjet parameters using regression analysis. *International Journal of Aerospace and Mechanical Engineering*, *5*(4), pp. 248–253.

Korat, M.M. and Acharya, G.D., 2016. A review on current research and development in abrasive waterjet machining. *International Journal of Engineering Research and Applications*, 4, pp. 432–423.

Kovacevic, R., Hashish, M., Mohan, R., Ramulu, M., Kim, T.J. and Geskin, E.S., 1997. State of the art of research and development in abrasive waterjet machining. *Journal of Manufacturing Science and Engineering*, *119*, pp. 776–785.

Kowsari, K., Sookhaklari, M.R., Nouraei, H., Papini, M. and Spelt, J.K., 2016. Hybrid erosive jet micro-milling of sintered ceramic wafers with and without copper-filled through-holes. *Journal of Materials Processing Technology*, *230*, pp. 198–210.

Kruusing, A., Leppaevuori, S., Uusimaki, A. and Uusimaki, M., 1999. Rapid prototyping of silicon structures by aid of laser and abrasive-jet machining. *Design, Test, and Microfabrication of MEMS and MOEMS*, *3680*, pp. 870–879.

Kumar, K.R., Sreebalaji, V.S. and Pridhar, T., 2018. Characterization and optimization of abrasive water jet machining parameters of aluminium/tungsten carbide composites. *Measurement*, *117*, pp. 57–66.

Kuriyagawa, T., Sakuyama, T., Syoji, K. and Onodera, H., 2001. A new device of abrasive jet machining and application to abrasive jet printer. *Key Engineering Materials*, 196, pp. 103–110.

Lin, Y.C., Hung, J.C., Lee, H.M., Wang, A.C. and Fan, S.F., 2018. Machining performances of electrical discharge machining combined with abrasive jet machining. *Procedia CIRP*, *68*, pp. 162–167.

Lu, Y., Tang, J., Ge, Z., Xia, B. and Liu, Y., 2013. Hard rock drilling technique with abrasive water jet assistance. *International Journal of Rock Mechanics and Mining Sciences*, *60*, pp. 47–56.

Melentiev, R. and Fang, F., 2018. Recent advances and challenges of abrasive jet machining. *CIRP Journal of Manufacturing Science and Technology*, *22*, pp. 1–20.

Molian, R., Neumann, C., Shrotriya, P. and Molian, P., 2008. Novel laser/water-jet hybrid manufacturing process for cutting ceramics. *Journal of Manufacturing Science and Engineering*, *130*(3), pp. 1–10.

Nanda, B.K., Mishra, A., Das, S.R. and Dhupal, D., 2020. Fluidized bed hot abrasive jet machining (FB-HAJM) of K-60 alumina ceramic. In *Advances in Unconventional Machining and Composites*, Editor: M. S. Shunmugam (pp. 641–650). Springer, Singapore.

Nguyen, T. and Wang, J., 2019. A review on the erosion mechanisms in abrasive waterjet micromachining of brittle materials. *International Journal of Extreme Manufacturing*, *1*, pp. 1–14.

Park, D.S., Cho, M.W., Lee, H. and Cho, W.S., 2004. Micro-grooving of glass using micro-abrasive jet machining. *Journal of Materials Processing Technology*, *146*(2), pp. 234–240.

Pawar, P.J. and Rao, R.V., 2013. Parameter optimization of machining processes using teaching–learning-based optimization algorithm. *The International Journal of Advanced Manufacturing Technology*, 67(5–8), pp. 995–1006.

Perec, A., 2016. Abrasive suspension water jet cutting optimization using orthogonal array design. *Procedia Engineering*, *149*, pp. 366–373.

Ramachandran, N. and Ramakrishnan, N., 1993. A review of abrasive jet machining. *Journal of Materials Processing Technology*, *39*(1–2), pp. 21–31.

Shafagh, S. and Papini, M., 2020. The effects of blast lag in abrasive jet machined micro-channel intersections. *Precision Engineering*, 62, pp. 162–169.

Sharma, V., Chattopadhyaya, S. and Hloch, S., 2011. Multi response optimization of process parameters based on Taguchi—Fuzzy model for coal cutting by water jet technology. *The International Journal of Advanced Manufacturing Technology*, *56*(9–12), pp. 1019–1025.

Tripathi, R., Srivastava, M., Hloch, S., Chattopadhyaya, S., Das, A.K., Pramanik, A., Klichová, D. and Adamcik, P., 2018. Performance analysis of pulsating water jet machining during disintegration of rocks by means of acoustic emission. In *Applications of Fluid Dynamics* (pp. 515–524). Springer, Singapore. Indian Institute of Technology (Indian School of Mines, Dhanbad, India.

Vanmore, V.V. and Dabade, U.A., 2018. Development of Laval nozzle for micro abrasive jet machining [MAJM] processes. *Procedia Manufacturing*, *20*, pp. 181–186.

Verma, A.P. and Lal, G.K., 1984. An experimental study of abrasive jet machining. *International Journal of Machine Tool Design and Research*, *24*(1), pp. 19–29.

Wakefield, C.M., Erichsen, G.A, Shuli, F.M., Stern. A.P., Naup, M., Meier, A., Jay, R. and Raghavan, C., 2008. Outline tracking device. Patent JP2008510969A.

Wakuda, M., Yamauchi, Y. and Kanzaki, S., 2003. Material response to particle impact during abrasive jet machining of alumina ceramics. *Journal of Materials Processing Technology*, *132*(1–3), pp. 177–183.

Wang, W., Biermann, D., Aßmuth, R., Arif, A.F.M. and Veldhuis, S.C., 2020. Effects on tool performance of cutting edge prepared by pressurized air wet abrasive jet machining (PAWAJM). *Journal of Materials Processing Technology*, *277*, p. 116456.

Yuzo, S. and Toi, Y., 2009. Method for processing arbitrary shape on workpiece made of conductive material and composite processing apparatus. Patent JP4288223B2.

Zain, A.M., Haron, H. and Sharif, S., 2011. Genetic algorithm and simulated annealing to estimate optimal process parameters of the abrasive waterjet machining. *Engineering with computers,* 27(3), pp. 251–259.

Zhang, L., Kuriyagawa, T., Yasutomi, Y. and Zhao, J., 2005. Investigation into micro abrasive intermittent jet machining. *International Journal of Machine Tools and Manufacture*, *45*(7–8), pp. 873–879.

Zhang, Y., Wang, Q., Hou, N. and Rao, S., 2020. Material removal mechanism of superalloy Inconel 718 based on electrochemical abrasive jet processing. *The International Journal of Advanced Manufacturing Technology*, *106*(11), pp. 1–11.

9 Enhancing Tribological Properties of Duplex Stainless Steel Via Electrical Discharge Treatment

A. Mahajan
Khalsa College of Engineering & Technology

S.S. Sidhu
Beant College of Engineering &Technology

S. Devgan
Khalsa College of Engineering & Technology

CONTENTS

9.1 INTRODUCTION

Duplex stainless steels (DSS) are a category of stainless steel that has high strength and excellent corrosion resistance with easy construction (Gregorutti et al. 2015). The metallurgical structure of DSS is a mixture of equal distribution of austenite and ferrite phases. Thus, DSS exhibit the characteristics of both austenite- and ferrite-type stainless steels (Lailatul and Abd Maleque 2018). DSS alloys consist of a higher content of chromium molybdenum, tungsten, nickel, and nitrogen. These alloys are significantly used in pollution control equipment, pressure vessels, transmission gears, heat exchangers, and piping in the chemical processing industry (Weibull 1987; Wang et al. 2012). However, the lower hardness and high wear rate of DSS constrained their utilization in tribological applications. The researcher observed that the wear resistance and hardness of steel surfaces can be enhanced by different surface modification techniques. Lailatul and Abd Maleque (2018) utilized the tungsten arc welding (TAW) torch technique to develop the SiC coating on the DSS surface. They reported that the resolidified layer has improved surface hardness (833.6 Hv) as compared to the untreated substrate (250 Hv). In order to develop a thin corrosive-wear resistance layer on the DSS 2205 substrate, Zhao et al. (2014) employed the anode plasma nitriding technique. They concluded that this surface modification technique enhances the substrate hardness and also increases the wear as well as corrosion resistance of DSS. However, Beloti et al. (2004) included the niobium in DSS composition and reported the improved biocompatibility as well as the mechanical properties of the DSS. Mourad et al. (2012) employed the laser beam welding technique for enhancing the corrosion resistance.

Surface treatment of various metals and alloys by electrical discharge machining (EDM) is highlighted by numerous researchers. This thermoelectrical technique can enhance the biological, physicochemical, and mechanical properties of metallic surfaces by depositing biocompatible layers (Mahajan et al. 2020; Devgan and Sidhu 2020a, b, c). Pramanik et al. (2018) utilized the wire EDM for the manufacturing of DSS. Mahajan and Sidhu (2019a, b) tailored the medical-grade Co-Cr alloy surface by EDM and reported that the machined surfaces show better corrosion resistance, biocompatibility, and wear resistance. However, Li et al. (2017) discussed the merits of powder-mixed electrical discharge machining (PMEDM) in the domain of biomaterials' surface modification. They reported that PMEDM ameliorates the wear performance, corrosion resistance, and bioactivity of the Ti alloy implant surface. Similarly, Devgan and Sidhu (2019) employed the PMEDM to develop the hydrophilic and crack-free surface on titanium beta-type alloy. Therefore, in this study, electrical discharge-treated DSS material is examined for its tribological performance. Herein, the pin-on-disk method is employed for investigating the wear rate and the coefficient of friction (COF) of the EDM-treated and unmachined DSS substrate.

9.2 MATERIALS AND METHODS

In this study, the square plate of DSS-2205 alloy (size of 30 mm length and 20 mm thickness) was supplied by Krone Impex (Mumbai, India). The chemical composition of the alloy is given in Table 9.1. The alloy plate was placed in deionized water (dielectric) tank, and then, the EDM process was carried out where a large amount of regular electric spark is discharged by a tool electrode that modifies the specimen surface. In this study, the machining of the DSS substrate was executed at optimum EDM parameters that were obtained from some past reported studies (Mahajan et al. 2019, 2020). The details about the processing parameters are presented in Table 9.2. After EDMing, the circular-shaped substrate (size: 12 mm diameter, 4 mm thickness) was removed from the square plate by wire EDM process. The substrate was further investigated for its tribological performance. The EDM setup is represented in Figure 9.1a.

TABLE 9.1
Chemical Composition of Chromium Cobalt (Co-Cr-Mo) Alloy

	Cr	Ni	Mo	Mg	Si	N	C	P	S	Fe
Base material	22%	5%	3%	2%	1%	0.2%	0.03%	0.03%	0.02%	Remainder

TABLE 9.2
Input Factor and Their Values

Input Factor	Values
Current (I), A	10
Pulse on-time (P-on), μs	60
Pulse off-time (P-off), μs	150
Workpiece	Duplex stainless steel (DSS-2205)
Electrode	Tungsten-Copper (W-Cu)
Dielectric medium	Deionized water

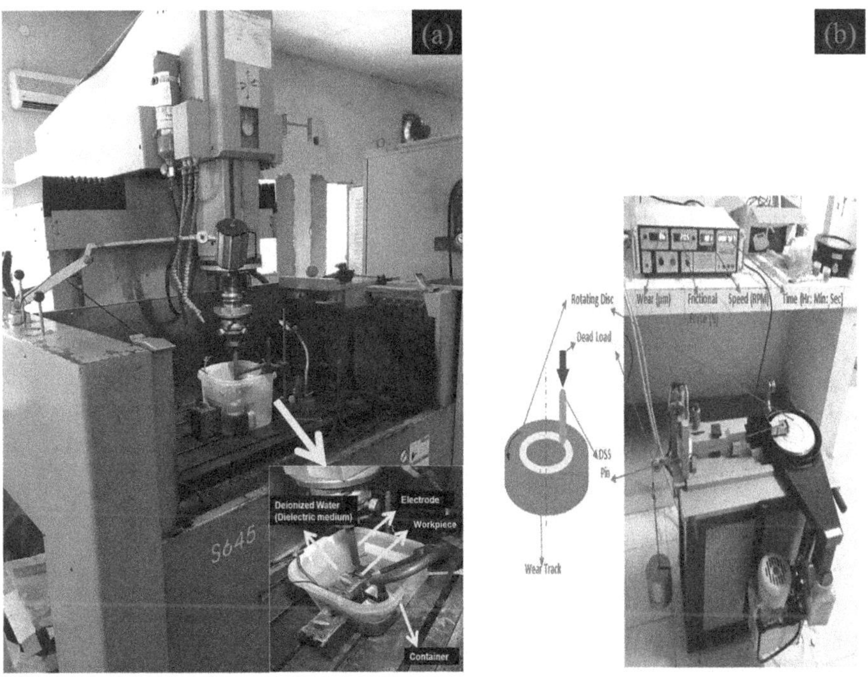

FIGURE 9.1 (a) Pictorial view of EDM setup and (b) pictorial and schematic view of pin-on-disk apparatus.

9.2.1 Pin-on-Disk Wears Studies

The wear test for the DSS substrate was carried out by utilizing a pin-on-disk-type tribometer supplied by Micromatic Technologies (Bangalore, India), as shown in Figure 9.1b. The untreated and EDMed substrates of DSS-2205 were constructed in the form of pin (40 mm length and 12 mm diameter), which were placed on the disk of EN31 steel (size of 120 mm diameter and 10 mm thickness). The ringer solution was used as a test lubricant, whereas the main wear track diameter for pins was considered as 80 mm. For each experiment, both pins were constantly loaded on the disk with a steady weight of 70N; spinning speed (disk) was 100 rpm, and the time duration was set at 1800 s. All experiments were carried out three times ($n = 3$) at ambient temperature. The control of the operation parameters and computation of wear and friction between disk and pin were acquired using computer and the TR-20LE software. However, from the substrates' mass loss technique, the materials' wear rates (W.R) in mm^3/Nm were calculated and expressed as:

$$W.R = \frac{\Delta m}{L \times \rho \times W} \tag{9.1}$$

where Δm represents the mass loss in grams, L is the total sliding distance in meters, ρ denotes the density of pin material (for DSS-2205: $\rho = 7.805$ g/mm^3), and W is the applied load in newton.

9.3 TRIBOLOGICAL INVESTIGATION OF EDMed SURFACES

Table 9.3 gives the wear performance and COF of both treated and untreated substrates. It has been witnessed that the amount of wear rate of an untreated substrate [$(3.52 \pm 0.15) \times 10^{-5}$] mm^3/N-m was more than the electrical discharge-treated substrate [$(1.05 \pm 0.08) \times 10^{-5}$] mm^3/N-m. Figure 9.2a shows the comparison of the wear rates of both substrates, whereas Figure 9.2b shows the COF with time for both pin samples. It has been depicted that the COF (μ) value of the untreated substrate ($\mu_{average} = 0.32$) is greater than that of the treated substrate ($\mu_{average} = 0.27$).

Wear rates for the EDM-treated and untreated samples were represented by field emission scanning electron microscopic (FESEM) images. The surface of the untreated substrate was found with flakes, scratches, and delaminations (Figure 9.3a).

TABLE 9.3
Calculated Wear Rate and COF of Substrates

		$N = 1$		$N = 2$		$N = 3$		Avg ± St. Dev.	
S. No.	Specimen Name	COF	Wear Rate (mm^3/m)	COF	Wear Rate (mm^3/m)	COF	Wear Rate (mm^3/m)	COF	Wear Rate (mm^3/m)
1	Untreated	0.317	3.52×10^{-5}	0.322	3.68×10^{-5}	0.312	3.37×10^{-5}	0.317 ± 0.005	$3.52 \times 10^{-5} \pm 0.15$
2	Treated	0.267	1.05×10^{-5}	0.276	1.13×10^{-5}	0.258	0.97×10^{-5}	0.269 ± 0.009	$1.05 \times 10^{-5} \pm 0.08$

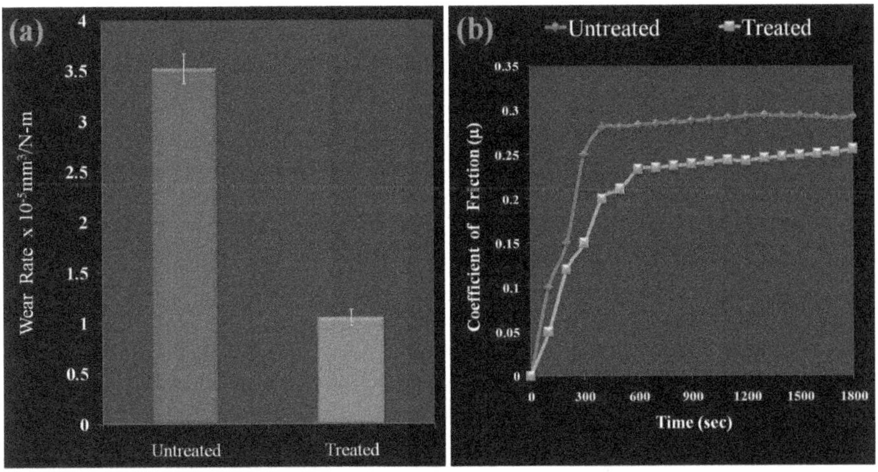

FIGURE 9.2 (a) Wear rate comparison of both substrates and (b) variation of the COF (μ) with a time of pin samples.

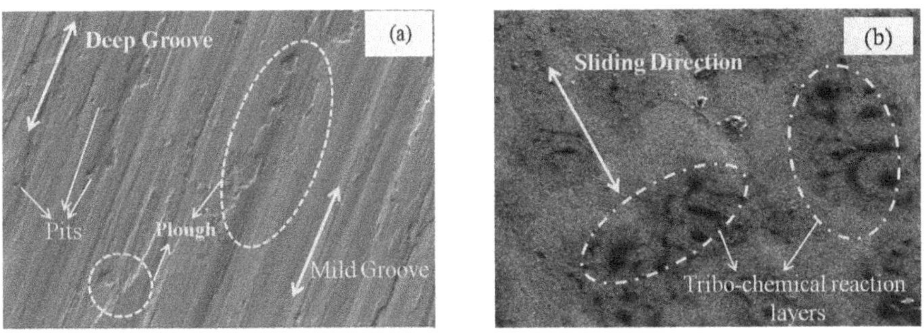

FIGURE 9.3 FESEM images of (a) untreated substrate and (b) treated substrate.

However, the EDMed substrate was witnessed with dark patches that represented the tribochemical reaction (wear of surface in the oxygen-rich atmosphere) (Figure 9.3b). It was also observed that the EDMed sample had lesser scratches and no delimitations that signify high wear resistance of the substrate (Ikeuchi et al. 2001; Hesketh et al. 2014). Evidently, during electrical discharge treatment (EDT), at high temperature, the chemical reaction between molten base elements and water (dielectric fluid) results in the formation of oxide and carbide layers on the surface, which could be the probable reason for the improved wear resistance of EDMed substrate. The results depicted improved wear resistance of EDM-treated substrate as compared to an untreated substrate, which results in the EDMed substrate to be used in various industrial applications.

9.4 CONCLUSIONS

This chapter focuses on the surface treatment of DSS-2205 by EDM technique. The pin-on-disk test was conducted to analyze the tribological performance. The results demonstrated that the EDMed surface has significantly higher wear resistance and a lower COF as compared to the untreated specimen. Therefore, the surface alteration via EDM can be deemed as a promising technique for enhancing the tribological properties of DSS substrates.

REFERENCES

Beloti, M. M., Rollo, J. M. D. D. A., Filho, A. I., & Rosa, A. L. (2004). In vitro biocompatibility of duplex stainless steel with and without 0.2% niobium. *Journal of Applied Biomaterials and Biomechanics*, *2*(3), 162–168.

Devgan, S., & Sidhu, S. S. (2019). Evolution of surface modification trends in bone related biomaterials: A review. *Materials Chemistry and Physics*, *233*, 68–78.

Devgan, S., & Sidhu, S. S. (2020a). Surface modification of β-type titanium with multi-walled CNTs/μ-HAp powder mixed electro discharge treatment process. *Materials Chemistry and Physics*, *239*, 122005.

Devgan, S., & Sidhu, S. S. (2020b). Enhancing tribological performance of β-titanium alloy using electrical discharge process. *Surface Innovations*, *8*(1–2), 115–126.

Devgan, S., & Sidhu, S. S. (2020c). Potential of electrical discharge treatment incorporating MWCNTs to enhance the corrosion performance of the β-titanium alloy. *Applied Physics A*, *126*(3), 1–16.

Gregorutti, R. W., Grau, J. E., Sives, F., & Elsner, C. I. (2015). Mechanical, electrochemical and magnetic behaviour of duplex stainless steel for biomedical applications. *Materials Science and Technology*, *31*(15), 1818–1824.

Hesketh, J., Ward, M., Dowson, D., & Neville, A. (2014). The composition of tribofilms produced on metal-on-metal hip bearings. *Biomaterials*, *35*(7), 2113–2119.

Ikeuchi, K., Morita, Y., Yoshida, H., & Kusaka, J. (2001). Effect of tribochemical reaction on wear of silicon carbide for joint prostheses. *Journal of Ceramic Processing Research*, *2*, 35–37.

Lailatul, P. H., & Abd Maleque, M. (2018). Tribological properties of surface coated duplex stainless steel containing SiC ceramic particles. *Journal Tribologi*, *18*, 136–148.

Li, L., Zhao, L., Li, Z. Y., Feng, L., & Bai, X. (2017). Surface characteristics of Ti-6Al-4V by SiC abrasive-mixed EDM with magnetic stirring. *Materials and Manufacturing Processes*, *32*(1), 83–86.

Mahajan, A., & Sidhu, S. S. (2019a). In vitro corrosion and hemocompatibility evaluation of electrical discharge treated cobalt–chromium implant. *Journal of Materials Research*, *34*(8), 1363–1370.

Mahajan, A., & Sidhu, S. S. (2019b). Potential of electrical discharge treatment to enhance the in vitro cytocompatibility and tribological performance of Co–Cr implant. *Journal of Materials Research*, *34*(16), 2837–2847.

Mahajan, A., Sidhu, S. S., & Ablyaz, T. (2019). EDM surface treatment: An enhanced biocompatible interface. In *Biomaterials in Orthopaedics and Bone Regeneration*, Editor: Preetkanwal Singh Bains (pp. 33–40). Springer, Singapore.

Mahajan, A., Sidhu, S. S., & Devgan, S. (2020). MRR and surface morphological analysis of electrical-discharge-machined Co–Cr alloy. *Emerging Materials Research*, *9*(1), 1–5.

Mourad, A. I., Khourshid, A., & Sharef, T. (2012). Gas tungsten arc and laser beam welding processes effects on duplex stainless steel 2205 properties. *Materials Science and Engineering: A*, *549*, 105–113.

Pramanik, A., Basak, A. K., Dixit, A. R., & Chattopadhyaya, S. (2018). Processing of duplex stainless steel by WEDM. *Materials and Manufacturing Processes*, *33*(14), 1559–1567.

Wang, Z. L., Wang, Z. X., & Zhang, Y. (2012). Study on welding parameters optimization of duplex stainless steel 2205 based on orthogonal. *Advanced Materials Research*, *472*, 1305–1308.

Weibull, I. (1987). Duplex stainless steels and their application, particularly in centrifugal separators: Part B Corrosion resistance. *Materials & Design*, *8*(2), 82–88.

Zhao, S., Wang, L., Xu, J. J., & Shan, Y. (2014). Surface modification of SS2205 duplex stainless steel by plasma nitriding at anodic potential. *Advanced Materials Research*, *881*, 1263–1267.

Index

Taylor & Francis Group
an **informa** business

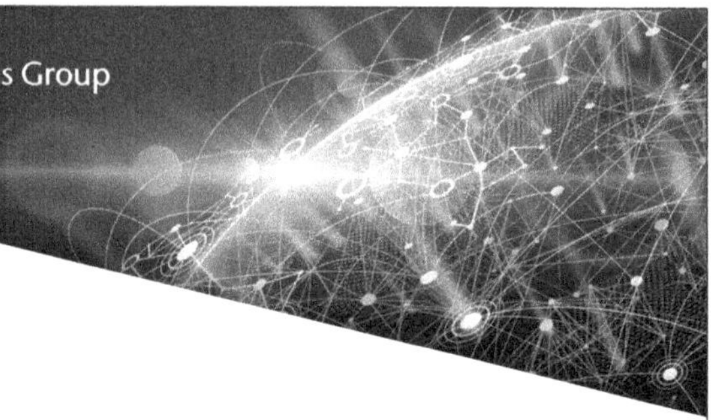

Taylor & Francis eBooks

www.taylorfrancis.com

A single destination for eBooks from Taylor & Francis with increased functionality and an improved user experience to meet the needs of our customers.

90,000+ eBooks of award-winning academic content in Humanities, Social Science, Science, Technology, Engineering, and Medical written by a global network of editors and authors.

TAYLOR & FRANCIS EBOOKS OFFERS:

- A streamlined experience for our library customers
- A single point of discovery for all of our eBook content
- Improved search and discovery of content at both book and chapter level

REQUEST A FREE TRIAL

support@taylorfrancis.com

Routledge
Taylor & Francis Group

CRC Press
Taylor & Francis Group